Undergraduate Topics in Computer Science

Undergraduate Topics in Computer Science (UTiCS) delivers high-quality instructional content for undergraduates studying in all areas of computing and information science. From core foundational and theoretical material to final-year topics and applications, UTiCS books take a fresh, concise, and modern approach and are ideal for self-study or for a one- or two-semester course. The texts are all authored by established experts in their fields, reviewed by an international advisory board, and contain numerous examples and problems. Many include fully worked solutions.

More information about this series at http://www.springer.com/series/7592

Joe Pitt-Francis · Jonathan Whiteley

Guide to Scientific Computing in C++

Second Edition

 Springer

Joe Pitt-Francis
University of Oxford
Oxford
UK

Jonathan Whiteley
University of Oxford
Oxford
UK

ISSN 1863-7310   ISSN 2197-1781   (electronic)
Undergraduate Topics in Computer Science
ISBN 978-3-319-73131-5   ISBN 978-3-319-73132-2   (eBook)
https://doi.org/10.1007/978-3-319-73132-2

Library of Congress Control Number: 2017962059

This Springer imprint is published by the registered company Springer International Publishing AG part
of Springer Nature
The registered company address is: Gewerbestrasse 11, 6330 Cham, Switzerland

Preface to the Second Edition

The principle changes in this updated edition are additional material on software testing and on some of the new features introduced in the C++11 standard. When introducing this additional material, we have followed the same philosophy as when writing the first edition of this book. That is, we focus on a concise discussion of the key features that are most useful to the novice and intermediate programmer in the field of scientific computing. We have found this an effective approach when teaching this course to graduate students—once the basics have been mastered, students then have the confidence to find out about less well-used features themselves when they are needed.

This second edition would not be as complete—or as enjoyable to update—without discussions with colleagues and other readers of the first edition, including those previously unknown to us who were kind enough to provide constructive feedback. We would like to express our gratitude to all who contributed in this way or offered their encouragement, and to the staff at Springer for inviting us to update the first edition.

Finally, we would both again like to thank our families for their love and support.

Oxford, UK Joe Pitt-Francis
October 2017 Jonathan Whiteley

Preface to the First Edition

Many books have been written on the C++ programming language, varying across a spectrum from the very practical to the very theoretical. This book certainly lies at the practical end of this spectrum and has a particular focus for the practical treatment of this language: scientific computing.

Traditionally, Fortran and MATLAB®[1] have been the languages of choice for scientific computing applications. The recent development of complex mathematical models—in fields as diverse as biology, finance and materials science, to name but a few—has driven a need for software packages that allow computational simulations based on these models. The complexity of the underlying models, together with the need to exchange code between co-workers, has motivated programmers to develop object-oriented code (often written in C++) for these simulation packages. The computational demands of these simulations may require software to be written for parallel computing facilities, typically using the Message Passing Interface (MPI). The need to train programmers in the skills to program applications such as these led to the development of a graduate-level course *C++ for Scientific Computing*, taught by the authors of this book, at the University of Oxford.

This book provides a guide to C++ programming in scientific computing. In contrast to many other books on C++, features of the language are demonstrated mainly using examples drawn from scientific computing. Object orientation is first mentioned in Chap. 1 where we briefly describe what this phrase—and other related terms such as inheritance—means, before postponing any further discussion of object orientation or related topics until Chap. 6. In the intervening chapters until object orientation reappears, we present what is best described as "procedural programming in C++", covering variables, flow of control, input and output, pointers (including dynamic allocation of memory), functions and reference variables. Armed with this grounding in C++, we then introduce classes in Chaps. 6 and 7. In these two chapters, where the main features of object orientation are showcased, we

---

[1]MATLAB is a registered trademark of The MathWorks, Inc.

initially, for the sake of clarity, abandon our principle of using examples drawn from scientific computing. Once the topics have been presented however, we resume our strategy of demonstrating concepts through scientific computing examples. More advanced C++ features such as templates and exceptions are introduced in Chaps. 8 and 9. Having introduced the features of C++ required for scientific computing, the remainder of the book focuses on the application of these features. In Chap. 10, we begin to develop a collection of classes for linear algebra calculations: these classes are then developed further in the exercises at the end of this chapter. Chapter 11 presents an introduction to parallel computing using MPI. Finally, in Chap. 12, we discuss how an object-oriented library for solving second-order differential equations may be constructed. The importance of a clear programming style to minimise the introduction of errors into code is stressed throughout the book.

This book is aimed at programmers of all levels of expertise who wish to write scientific computing programs in C++. Experience with a computer to the level where files can be stored and edited is expected. A basic knowledge of mathematics, such as operations between vectors and matrices, and the Newton–Raphson method for finding the roots of nonlinear equations would be an advantage.

The material presented here has been enhanced significantly by discussions about C++ with colleagues, too numerous to list here, in the Department of Computer Science at the University of Oxford. A special mention must, however, be made of the Chaste[2] programming team: particular gratitude should be expressed to Jonathan Cooper for readily sharing with us his impressively wide and deep knowledge of the C++ language. Other members of the team who have significantly helped clarify our thoughts on the C++ language are Miguel Bernabeu, James Osborne, Pras Pathmanathan and James Southern. We should also thank students from both the M.Sc. in Mathematical Modelling and Scientific Computing and the Doctoral Training Centres at the University of Oxford for unwittingly aiding our understanding of the language through asking pertinent questions.

Finally, it is always important to remember—especially when debugging a particularly tiresome code—that there is far more to life than C++ programming for scientific computing. We would both like to thank our families for their love and support, especially during the writing of this book.

Oxford   Joe Pitt-Francis
October 2011   Jonathan Whiteley

---

[2]The Cancer, Heart And Soft Tissue Environment (Chaste) is an object-oriented package, written in C++, for simulations in the field of biology. More details on this package may be found at https://www.cs.ox.ac.uk/chaste/.

Contents

Getting Started

1

In this introductory chapter, you will learn a little bit about the features of C++ in terms of some of the common "buzzwords" you may have heard about the language, and in terms of its strengths and weaknesses. You will also learn how to edit, compile and run your first C++ program. This chapter also includes information on variables and simple ways of getting data into and out of your programs.

The chapter concludes with tips on how you might, as a novice C++ programmer, go about debugging your programs. We have included tips with every chapter in this book. They are presented at an increasing level of sophistication—this should match your gaining knowledge as you read through the book and attempt some of the exercises.

1.1 A Brief Introduction to C++

A very large number of programming languages for writing computer software exist. If one of these programming languages was the most suitable for all purposes, then it would be expected that everyone would use this language, and all other languages would eventually become obsolete. This, however, is certainly not the case. It seems appropriate to begin this book by describing the key features of C++, allowing us to explain why C++ is a suitable programming language for scientific computing applications and why it isn't the only suitable choice of language.

© Springer International Publishing AG, part of Springer Nature 2017
J. Pitt-Francis and J. Whiteley, *Guide to Scientific Computing in C++*, Undergraduate Topics in Computer Science, https://doi.org/10.1007/978-3-319-73132-2_1

1.1.1 C++ is "Object-Oriented"

You may have heard that C++ is an "object-oriented" language and have wondered what that means. What marks a language which is object-oriented out from one that is not? Fundamentally, it is because the basic unit of the language is an *object* or *class*—an entity which brings together related functionality and data. We will probe the ideas behind objects and classes more deeply in Chap. 6.

Many books on C++ start by defining *object-orientation* more explicitly. If this book were aimed at a computer science or software engineering audience, then we would find it necessary to define some specific concepts related to object-orientation. We would need to convince you of the importance of the following concepts.

- *Modularity*. All the data of a particular object, and the operations that we perform on this object, are held in one or two files, and can be worked on independently.
- *Abstraction*. The essential features and functionality of a class are put in one place and the details of how they work are unimportant to the user of the class. For example, if you are using a linear system library to solve matrix equations you should not need to know the precise details of how matrices are laid out in memory or the exact order that a numerical solver performs its operations. You should only need to know how to use the functionality of the library.
- *Encapsulation*. The implementation of an object is kept hidden from the user of the class. This is not only about clarity (*abstracting* away the detail). It is also about preventing the user from accidentally amending internal workings of, for example, a linear solver, stopping it from working effectively.
- *Extensibility*. Functionality can be reused with selected parts extended. For example, much of the core of a linear solver is in matrix-vector products and scalar products—this type of functionality need only be implemented once, then other parts of the program can build on it.
- *Polymorphism*. The same code can be used for a variety of objects. For example, we would like to use similar looking C++ code to raise a matrix of complex numbers to a given power as we would to raise a real number to a given power—even though the basic arithmetic operations "behind the scenes" are different.
- *Inheritance*. This, perhaps the most important feature of object-orientation, allows for code reuse, extensibility and polymorphism. For example, a new linear solver for singular matrix systems will share many of the features of a basic linear solver. Inheritance allows the new solver to derive functionality from the basic solver, and then build on this functionality.

We are not going to discuss these terms in any more detail at this time. It is not that these things are unimportant. Quite the contrary—all these concepts add up to make C++ a very powerful language. However, we can cover the basics of programming without object-orientation. We will describe classes and objects in Chap. 6 and revisit some of these concepts. Then we can show exactly why *inheritance*, for instance, is so powerful when we come to explain it in Chap. 7.

1.1.2   Why You Should Write Scientific Programs in C++

Since you have selected a book with the words "C++" and "Scientific Computing" in the title, then the chances are that you have decided to start writing your scientific programs in C++. Perhaps not. Perhaps you are considering your options, or perhaps the choice of language has been foisted on you.

It is not our place to fight battles about which language is the very best, especially because the choice of language for a program will often depend on the problem that is being solved. In the field of numerical scientific programming, there are many languages being used, with most scientists opting for MATLAB®,[1] C/C++ or Fortran.

The first and most compelling reason for using C++ (as well as C and Fortran) is because they are *fast*. That is, with careful programming and optimisations, they can be compiled to a machine code program which is able to use the full power of the available hardware. Many scripting languages (such as MATLAB and Python) are *interpreted* languages, meaning that the code which you write is translated to machine code at run time. Other modern languages (such as Java and C#) compile halfway—to a hardware-independent byte-code which is then interpreted at run time. Run time interpretation means that some of the computer's power is spent on the conversion process and also that it is harder to apply optimisations. Nowadays MATLAB, Python and Java implementations use clever tricks such as caching compilation steps and just-in-time compilation to make programs run faster. Nevertheless, these tricks require computational effort and so these languages may not fully utilise the power of all hardware.

A second reason for using C++ is that there is a *wealth of numerical libraries* for scientific computing in C++ and related languages. Lots of numerical algorithms were established in the 1950s and were then incorporated into software libraries (such as EISPACK and LINPACK) in the 1970s.[2] If you write your own code using well-established, well-tested software then you are building on decades of experience and improvement.

A third reason for choosing to write in C++ is that there is a *wide-range of open source and commercial tools* to support you. We used the free GNU compiler tool-set to test the programs in this book and you can use any C++ compiler to compile them for your computer. In contrast, if we were distributing MATLAB programs, you would need to have MATLAB and a licence installed on your computer because it is a proprietary product. There are similar open source products (such as GNU Octave) but there is no guarantee that a MATLAB program will produce the same answer when run in Octave. Because it is closed source, the meaning of a program can change between versions of MATLAB. For example, when just-in-time compilation was introduced in MATLAB 7 the operational semantics of the language subtly changed. This meant that a small minority of MATLAB programs which were known to work well with one

---

[1] MATLAB is a registered trademark of The MathWorks, Inc.
[2] The original version of MATLAB was written in Fortran and was intended as a simple interface into parts of the EISPACK and LINPACK Fortran libraries.

version of MATLAB could produce incorrect results, errors or warnings on another version.

A fourth reason for C++ is that it has a *flexible memory management model*. In a Java program, some of the system memory is used in the interpretation and you rely on a garbage collector to tidy up memory which you are no longer using, and so you may not be able to predict how much memory a program is going to need. In C++ you can make this prediction, but this is a double-edged sword because you are also responsible for making sure that memory is managed properly.

A final reason to program in C++ is that it is an *object-oriented language*. We haven't yet told you what this means exactly, but it is widely held that writing in an object-oriented style leads to programs which are easier to understand, to extend, to maintain and to refactor.

1.1.3 Why You Should Not Write Scientific Programs in C++

It is worth stressing that C++ is not the best language for every occasion. Some people say that *other languages may be faster*. Many scientific programmers believe that Fortran will always give the best performance in terms of raw speed and would reject C++ on the basis that features such as pointer chasing and virtual method look-up (don't worry if you haven't heard of these terms, or don't know what they mean—you may never need to!) result in the code being executed at suboptimal speed. This may have some truth, but the fact that object-orientation leads to greater readability (as mentioned above) makes it a reasonable compromise language. It can be a very fast language and it is also a good language for readability.

Sometimes *other languages are better for a specialised task*. Scripting languages such as Perl and Python are ideal for text processing and string manipulation. If you need to sum columns of numbers from files then you could write a C++ program, but a short, disposable script would be far quicker to implement.

Some languages are *better for writing prototype programs or for plotting data*. MATLAB excels in the field of rapid prototyping—short programs to quickly explore some algorithm or phenomenon. To test a particular linear algebra algorithm on a range of matrices with various sizes and structures would take a few lines of MATLAB, but in C++ you might have to write several files and compile against someone else's libraries. MATLAB also has the advantage of a fully-integrated graphical development environment, making many programming tasks easy without having to rely on extra tools. Furthermore, MATLAB has an in-built plotting environment, so if you want to visualise the results of your algorithms quickly MATLAB might be your best choice.

So C++ may not be the best choice of language in *every* situation. However, there are many situations in which C++ has the ideal fit for a particular problem. The discussion above may be enough to convince you that it is worth getting started with C++.

1.1.4 Scope of This Book

Most C++ programs for scientific computing can be written very effectively by using only a fraction of the total capabilities of the language. This book focuses on the aspects of C++ that you are most likely to utilise, or to encounter in other programmer's code, for scientific computing applications. When writing your own programs, you may occasionally need to understand one of the more advanced features of the language. In the Further Reading section at the end of this book, we direct the reader to a collection of resources that provide a more comprehensive description of the whole C++ language [5–8].

1.2 A First C++ Program

It is very common to introduce a programming language by using a program that prints the text "Hello World" to the screen. A simple example of a C++ program that does this is shown below. The code in Listing 1.1 illustrates several basic features of C++ programs. In line 1 of this code, we include the header file `iostream`. The name `iostream` pertains to input and output **stream**ing and is required in any C++ program that inputs data from the keyboard or outputs data to the console, that is, the screen. The second feature to note is that there is a section of code that:

- begins with the line of code "`int main(int argc, char* argv[])`" (line 3 of this code);
- is followed by more code enclosed between curly brackets, { and }; and
- the code within the curly brackets ends with the statement "`return 0;`".

The section of the code between curly brackets contains the instructions that we want the computer to execute. The part of line 3 inside brackets allows us to execute the code using user-specified arguments: we will postpone a discussion of this functionality until Chap. 3. Note that comments have been inserted into the code in lines 5, 6, 7 and 9 to aid the reading of the code by humans: anything between the comment opener "`/*`" and the comment closer "`*/`", or any line that starts with "`//`" is a comment, and is ignored when the code is converted into an executable, computer readable file. We have used the extension `.cpp` for the code below to indicate that the file `HelloWorld.cpp` is a C++ program. Choice of this extension is entirely a matter of personal choice: other authors use the extensions `.C`, `.c++`, `.cxx` or `.cc`.

We now focus on the purpose of lines 10 and 12: these lines of code each contain an instruction to the computer, and are known as *statements*. Note that all statements end with a semi-colon. It is sufficient for the time being for the reader to know that line 10 is the line of code that directs the computer to print the contents within the quotation marks to the screen. The "`\n`" denotes a new line, and so the phrase "Hello World", followed by a new line, will be printed to the screen. The word `cout` is a contraction of **c**onsole **out**put, that is, printing to the screen.

Listing 1.1 HelloWorld.cpp

```
1  #include <iostream>
2
3  int main(int argc, char* argv[])
4  {
5      /* This is a comment and will be ignored by the compiler
6      Comments are useful to explain in English what
7      the program does */
8
9      // Print "Hello World" to the screen
10     std::cout << "Hello World\n";
11
12     return 0;
13  }
```

The word "int" at the start of line 3 indicates that the last line of the code within curly brackets will return an integer value. This is carried out using the statement in line 12 "return 0;". Returning the value zero indicates to the computer that the program has reached the end without encountering any problems.

Before moving on to explain how to get your computer to print Hello World to your screen we pause to discuss some stylistic issues of which you should be aware. You will see in the listing above that all lines of code within the curly brackets have been indented. This is not compulsory. However, it is standard practice when coding to indent these lines: this will become clearer in later chapters when we embed code within more than one set of curly brackets. The number of spaces indented is entirely for the programmer to decide: all spaces—termed "white space"—are ignored when executing the code above. A final point is that lines in C++ may be as long as the programmer wishes, and may run over the end of the line in the text editor used to write your C++ programs. For clarity, it is generally advisable to split a potentially long line over several lines. We will demonstrate this later when writing more complex statements.

The code in Listing 1.1 is a correct C++ program for printing the text "Hello World" to the screen. However, before this program may be executed it must first be translated into a format that the computer can read: this process is known as *compilation*. We now explain what compilation is, and how to do it.

1.3 Compiling a C++ Program

Many readers will have experience of scientific computing in MATLAB. A key difference between C++ and MATLAB is that a C++ program must be *compiled* before it can be executed. There are many different ways that compilation can be performed which we now discuss.

1.3.1 Integrated Development Environments

As you take your first steps in learning a new programming language, you may not want to invest a lot of time in installing new software and configuring applications to help you develop programs. For this reason, we recommend that you begin writing programs with your favourite text editor and a command line compiler (see the following Sect. 1.3.2). However, as your programs and projects grow in size you will need to manage multiple files each containing various parts of the program. This becomes difficult when the number of files becomes large, and you may spend a lot of time switching between files in order to look up what you called some function or argument. At this point in your code development, we would recommend that you switch to using an Integrated Development Environment (IDE).

Examples of IDEs that are available for C++ programmers at the time of writing include KDevelop for Linux, Microsoft Visual Studio for Windows, XCode for Mac OS X, and the cross-platform IDEs CLion and Eclipse. Eclipse is open source, runs on most operating systems and is well-maintained by a community of developers. Because it was originally built for developing Java programs, it is necessary to install a "C/C++ development tools plug-in" should it be used for developing C++ programs.

The functionality of various IDEs varies according to their level of sophistication, but most present the seasoned programmer with several advantages over an old-school compile at the command line approach. Common features of IDEs are listed below. Don't worry if you do not fully understand all the terms used: these will become clear as you work through this book.

1. A program editor with syntax highlighting such as keyword colouring, automatic code indentation and identification of illegal programming constructs.
2. Context aware editing, so that you immediately know what functionality is present in one of your classes as you type its name.
3. Build automation, where your entire project code is managed so that changes to small parts of a large program only result in small compilation steps. Build automation is traditionally done with a hand-crafted file known as a `Makefile`, which we introduce in Sect. 6.2.4.1. Many IDEs analyse your code for dependencies and then use a `Makefile` behind the scenes.
4. On-the-fly compilation gives the system the ability to constantly save and compile your program as you write it.
5. "Step through" graphical debugging lets you walk through a program as it runs, pause it at critical points, and examine the internal state of its variables. (More information on debuggers is given in Sect. 7.7.)
6. Automatic code generation is particularly useful in IDEs for graphical tool development. When the user selects that they want to include a button on a graphical tool in their program some "boiler plate" code is generated including the functions that are activated when the button is pressed—these are then filled in by the programmer.

1.3.2   Compiling at the Command Line

When using the Linux operating system,[3] C++ codes may be compiled and executed at the command line within a terminal window. Many compilers—both open source and commercially developed—are available. In this book, we assume that the reader has access to the GNU `gcc` compiler. To ensure that this compiler is installed, open a terminal window and type "`which g++`" followed by return. Hopefully the computer will respond by reporting the location of this compiler, for example,

```
1  $ which g++
2  /usr/bin/g++
3  $
```

If the compiler is not installed, it may be downloaded from https://gcc.gnu.org/, where instructions for installation may also be found.

To compile the code given in Listing 1.1, open a terminal window and create a directory where code may be saved. Move into this directory, and save the code as "`HelloWorld.cpp`". In the same directory type

```
g++ -o HelloWorld HelloWorld.cpp
```

In the command above, `g++` tells the computer that we want to use the GNU `gcc` compiler for C++. The section of the command "`-o HelloWorld`" tells the computer that we want to name the executable file "`HelloWorld`". The "`-o`" is known as the *flag* that the computer expects will be followed by the executable name, in this case `HelloWorld`. The command ends by stating the C++ file that we wish to compile. This command produces an executable file called `HelloWorld`. This executable may be run by typing "`./HelloWorld`" inside the terminal. Running this executable will result in the text "Hello World" being printed to the screen inside the terminal.

If we were to compile the code using the command above, but without the flag and the executable name, then an executable file would still be produced. A default name would be allocated to the executable file. For many compilers, this default executable name is `a.out`.

---

[3]If you are working on a Mac operating system, we recommend that you install the Xcode developer tool-set. This comes complete with a GNU C++ compiler which you can use on the command line or within the developer environment. If you are working on a Windows operating system, we recommend that you install MinGW (a minimal environment for using GNU tools within Windows). Alternatively, you may want something more sophisticated built on MinGW such as Cygwin (a Unix-like environment) or Code::Blocks (an open source windows development environment containing MinGW and the GNU C++ compiler).

1.3.3 Compiler Flags

If we were to attempt to compile a code that was not written using correct C++ syntax, then the compiler would report an error, and would not produce an executable file. As such, the compiler can be thought of as a helpful tool that has the capability to perform some validation of the correctness of the code.

Suppose we have written code where a calculation was stored as a variable, but this variable is never subsequently used. Although this may be written with correct C++ syntax it is likely that this is an error—we would expect that the result of every calculation will subsequently be used somewhere in the code, or there would be no point in performing this calculation. Compilers have the capacity to warn us of unexpected occurrences such as this by the use of *compiler flags*. The compilation command below will warn us of instances such as these.

```
g++ -Wall -o HelloWorld HelloWorld.cpp
```

The compiler flag -Wall above is a contraction of **warning all**. The compilation command above will warn us of anything unexpected that is not actually an error, but will still create an executable file. We give an example instance of a situation in which the compiler will warn of a probable programming error as one of our programming tips in Sect. 2.6.3. Suppose we want to be stricter than this, and want the compiler to treat anything unexpected as an error and, therefore, not to create an executable file when this occurs. This may be achieved using the compilation command below.

```
g++ -Wall -Werror -o HelloWorld HelloWorld.cpp
```

There are a large number of compiler flags available for most compilers. At this stage, there is no need to know about any more than the basic flags. We have shown how to use compiler flags to perform some validation of the code written. We will now discuss three more flags that are particularly valuable when writing scientific computing applications. The first flag we discuss may be used to optimise the performance of the executable file. The default is no optimisation. By using the "-O" (upper case o) flag as shown below, the executable file should execute more quickly although compilation may take longer.

```
g++ -O -o HelloWorld HelloWorld.cpp
```

If we are debugging a program, it is important that the executable and the debugger have information about which line in the source code produced specific machine instructions. Normally this information is not retained after compilation. In order to produce a non-optimised version of the code with debugging information preserved, we use the "-g" flag.

```
g++ -g -o HelloWorld HelloWorld.cpp
```

The last flag that we introduce here is one that allows us to link to a library of mathematical routines. We instruct the compiler to link to this library using the command below.

```
g++ -lm -o HelloWorld HelloWorld.cpp
```

We may use as many flags as we wish when compiling—simply list them one after the other when compiling the code.

1.4 Variables

In the example code in Listing 1.1 we simply printed some text to the screen. In most programs, especially scientific computing applications, we wish to store entities and perform operations on them. These entities are known as *variables*. In C++ programs, in common with most compiled languages, the variables must be declared to be an appropriate type before they are used.

1.4.1 Basic Numerical Variables

The two most common types of variable that are used in scientific computing applications are *integers* and *double precision* floating point variables. Loosely speaking, if a numerical variable does not—and never will—require a decimal point it may be stored as an integer variable: if not it should be stored as a floating point variable. If a code uses two integers denoted by row and column, and one double precision floating point variable denoted by temperature, we may declare these before they are used, and set their values, using the following code fragment.

Listing 1.2 Declaring variables

```
1    int row, column;
2    double temperature;
3    row = 1;
4    column = 2;
5    temperature = 3.0;
```

The statements in lines 1 and 2 of the code above allocate memory for two integer variables `row` and `column`, and one double precision floating point variable `temperature`. It is important to understand that, whilst memory is allocated for these variables, we do not know until we assign values to these variables in lines 3–5 what values are stored by these variables. A common mistake is to assume that these variables are initialised to zero when the memory is allocated: this is true some of the time, but you should not rely on this.

Note the use of the decimal point for the double precision floating point variable `temperature` in line 5 of the listing above. This is not strictly necessary, but emphasises that this variable is a floating point variable. Use of this decimal point has the advantage that, provided we compile the code with suitable flags, compilation will trigger a warning if we had mistakenly declared this variable to be an integer.

We strongly encourage the use of variable names that have some relation to the variable that they represent, for example `row` as a variable that contains the index to the row of a matrix (see Sect. 6.6 for a longer discussion of naming conventions for variables). There are certain rules that variable names in C++ must adhere to, but these rules are not particularly restrictive. The first rule is that all variables in C++ programs should begin with a letter. All other characters in variable names must be letters, numbers or underscores. Variable names are case–sensitive, and so "ROW" is a different variable to "`row`". We would not, however, recommend writing a program with one variable called "ROW" and another variable called "`row`" as the potential for confusing these variables is obvious. One final restriction is that some names, such as `int, for, return` may not be used as variable names because they are used by the language. These words are known as *reserved words* or *keywords*.

A variable may be *initialised* when defining the variable type. For example, the code fragment in Listing 1.2 may be written as the following code fragment.

```
int row = 1, column = 2;
double temperature = 3.0;
```

The value of more than one variable may be assigned in each statement, as shown below.

```
int row = 1, column = 2;
row = column = 3;
```

However, line 2 in the code fragment above may cause confusion—it actually means

```
int row = 1, column = 2;
row = ( column = 3 );
```

and so both `row` and `column` take the value 3 after this fragment of code has been executed. However, it may be mistakenly read to be

```
int row = 1, column = 2;
( row = column ) = 3;
```

in which `row` would first take the value 2 (which was the initial value of column), and then `row`, because it is the *result* of the assignment `row = column`, would take the value 3. The value of column is unaffected. There is clearly potential for introducing errors when assigning more than one value in each statement, and so we do not recommend this approach.

It is often the case that a programmer intends a variable to be constant throughout the code, for example the numerical value used for the density of a fluid. The programmer can ensure that a variable is guaranteed to be unchanged throughout the code by assigning a value to the variable when it is declared, together with use of the keyword `const` as shown in the fragment of code below.

```
const double density = 45.621;
```

We may want to set the tolerance of some iterative solver to a very small number, for example 10^{-12}. Clearly, we may set this tolerance using the code fragment below.

```
double tolerance = 0.000000000001;
```

The listing above is clearly not ideal—a casual glance at the code does not allow us to distinguish easily between, say, 10^{-10} and 10^{-12}. It would be much clearer if we could write the numerical value in *scientific notation*. This is demonstrated in the code below.

```
double tolerance = 1.0e-12;
```

The letter "e" in the line of code above may be read as "times 10 to the power of": that is, 589.63 may be written `5.8963e2` as $589.63 = 5.8963 \times 10^2$.

1.4.2 Other Numerical Variables

In the previous section, we restricted ourselves to declaring all integer variables using the keyword `int` and all floating point variables using the keyword `double`. There are—however—variants on these variable types which we now discuss.

Integers can be declared as *integers*, *short integers* or *long integers* as shown below.

```
int integer1;
short int integer2;
long int integer3;
```

The actual range of integers that may be stored by each of these variables depends on the system that you are using. For example, on an obsolete 32-bit operating system the long int is completely synonymous with the int data type—but on modern 64-bit architectures the long int is assigned twice as much space as the int (so it can store numbers in the range $\pm 9 \times 10^{18}$ as opposed to $\pm 2 \times 10^9$).

Variables of type short int require the allocation of less memory, with a corresponding reduction in the range of values that may be stored in this memory. It may be tempting to try to use short integers where possible to free up as much memory as possible. We do not recommend this: in software written for scientific computing applications the bulk of memory allocated is usually used to store floating point variables. Reducing the memory allocated to integer variables is unlikely to free a significant volume of memory.

A further classification of each of the integer types is as *signed* or *unsigned* integers. Signed integers may be used to store both positive and negative integers, whilst unsigned integers may be used to store only nonnegative integers. These variables may be used as shown below.

```
signed long int integer4; // signed is unnecessary
unsigned int integer5;
```

The default for any integer is a signed integer, hence there is no purpose in explicitly declaring an integer as a signed integer. A variable of type unsigned int is allocated an identically sized memory location as a variable of type int. As would be expected, a variable of type unsigned int can then store a range of nonnegative integers roughly twice as big as a variable of type int. A programmer is, however, unlikely to notice the difference between these two variable types on modern systems.

Floating point variables may be declared using the keywords float, double or long double as shown below.

```
float floating_point_number1;
double floating_point_number2;
long double floating_point_number3;
```

As with integers, the range of numbers that may be stored using each of these variable types depends on the system used. On modern systems it is very rare that the range of numbers that may be stored by a variable of type `double` differs from the range that may be stored by a variable of type `long double`. In the remainder of this book, we do not distinguish between these data types. Variables of type `float` typically store a smaller range of numbers than those of type `double`. Although variables of type `double` require more memory we strongly urge writers of scientific computing applications to use double precision floating point variables: this will minimise the effect of rounding errors, thus removing one potential source of error from any program written.

1.4.3 Mathematical Operations on Numerical Variables

Sample C++ code for performing a variety of mathematical operations on variables is given below. Note the inclusion of the header file `cmath`. This file is needed for some mathematical operations and also includes values of some useful constants, such as `M_PI`, that contains the value of π correct to about 20 decimal places.

```
1   #include <cmath>
2
3   int main(int argc, char* argv[])
4   {
5       double x = 1.0, y = 2.0, z;
6       z = x/y;           // division
7       z = x*y;           // multiplication
8       z = sqrt(x);       // square root
9       z = exp(y);        // exponential function
10      z = pow(x, y);     // x to the power of y
11      z = M_PI;          // z stores the value of pi
12
13      return 0;
14  }
```

Many other mathematical functions are available. The functions `cos`, `sin`, `tan`, `acos`, `asin`, `atan`, `cosh`, `sinh`, `tanh`, `log`, `log10`, `ceil`, `floor` can be used in exactly the same way as `sqrt` and `exp` in the code above: that is, they accept one argument, and return one value.

Some mathematical functions deserve more explanation. This is done through their implementation in code below.

```cpp
#include <cmath>

int main(int argc, char* argv[])
{
  double x = 7.8, y = 1.65, u = -3.4, z;
  z = fmod(x, y);    // remainder when x is divided by y
                     // z is 1.2 since 7.8 = 4*1.65 + 1.2
  z = atan2(y, x);   // inverse tangent (in radians) of
                     // angle between the vector
                     // (x, y) and the positive x-axis
                     // note the ordering of y and x in
                     // calling the function atan2
                     // z is 0.208465
  z = fabs(u);       // Absolute value of u
                     // z is 3.4
                     // note fabs should not be confused
                     // with abs (the integer equivalent)

  return 0;
}
```

There are many instances in scientific computing code where we wish to increment a variable a by the value b, that is, we want to replace the value that the variable a stores by the value a+b. There are shorthand operations for this and other similar operations in C++, shown in Table 1.1.[4] Note that the a%b operation, pronounced "a mod b", is a modulus operation and may be thought of as the remainder after dividing a by b using integer division as described in Sect. 1.4.4.

Table 1.1 Shorthand for some mathematical operations

Longhand	Shorthand
a = a + b;	a += b;
a = a - b;	a -= b;
a = a * b;	a *= b;
a = a / b;	a /= b;
a = a % b;	a %= b; if a and b are integers ($a \bmod b$)
a = a + 1;	a++; if a is an integer
a = a - 1;	a--; if a is an integer

---

[4]The "++" shorthand programming construct, which is also available in the C language, explains the original naming of the language "C++". It is a pun which means "like C but one better".

1.4.4 Division of Integers

One common error frequently made by inexperienced C++ programmers is in dividing an integer by another integer. Consider the fragment of code below.

```
1    int i = 5, j = 2, k;
2    k = i / j;
3    std::cout << k << "\n";
```

This code fragment will output the value 2, when the value of dividing 5 by 2—that is, 2.5—was actually intended. There are two potential problems with the code fragment as it is written above. The first operation that will be performed when executing line 2 of the listing above is to divide the integer i by the integer j. The value resulting from this operation will then be stored in the memory allocated to k. In C++, division of an integer by another integer will return *only the integer part of this division*: hence dividing i by j will store the integer part of 2.5, which is 2 (as everything after the decimal point will be ignored). The second part of this statement—the assignment operator—will then assign the value 2 to the integer variable k.

It may be thought that modifying the code fragment above so that k is defined to be a double precision floating point variable may solve the problem, as shown in the code fragment below.

```
1    int i = 5, j = 2;
2    double k;
3    k = i / j;
4    std::cout << k << "\n";
```

This still does not give the correct value of 2.5. This is because the division is performed in line 3 before the result is stored as the double precision floating point variable k. As division of an integer by another integer in C++ returns the integer part of the division, the division of i by j returns the value 2 as explained above. This value is then stored as the double precision floating point number 2.0 in the memory allocated to k.

To divide two integers as if they were floating point variables, we may convert the integers to double precision floating point variables as shown in the code fragment below.

```
1    int i = 5, j = 2;
2    double k;
3    k = ((double)(i)) / ((double)(j));
4    std::cout << k << "\n";
```

The code `((double)(i))` is known as *"explicit type conversion"* and allows us to treat the integer variable `i` as a double precision floating point variable, and so this code fragment does output the correct value of 2.5.

1.4.5 Arrays

Many scientific computing applications are underpinned by algorithms that are based on vectors and matrices. These may be stored in C++ as an entity known as an *array*. If the size of the array is known in advance then it can be declared as follows.

```
int array1[2];
double array2[2][3];
```

In the code fragment above, `array1` represents a vector of integers of length 2, whilst `array2` represents a matrix of double precision floating point variables of size 2×3.

In contrast to MATLAB and Fortran, in C++ the indices of an array of length n start with entry 0 and end with entry n-1. This is known as *"zero-based indexing"*. Elements of an array are accessed by placing the indices in separate square brackets, and so we may completely populate the arrays `array1` and `array2` declared above using the following code.

```
array1[0] = 1; // Note that indexing begins from 0
array1[1] = 10;
array2[0][0] = 6.4;
array2[0][1] = -3.1;
array2[0][2] = 55.0;
array2[1][0] = 63.0;
array2[1][1] = -100.9;
array2[1][2] = 50.8;
```

We may also perform operations on entries of the array as shown below.

```
array1[0]++; // increments the value of this entry by 1
array2[1][2] = array2[0][1] + array2[1][0];
```

Arrays can be initialised when they are declared, for example,

```
double array3[3] = {5.0, 1.0, 2.0};
int array4[2][3] = { {1, 6, -4}, {2, 2, 2} };
```

where the array `array3` represents the vector

$$\begin{pmatrix} 5 \\ 1 \\ 2 \end{pmatrix},$$

and `array4` represents the matrix

$$\begin{pmatrix} 1 & 6 & -4 \\ 2 & 2 & 2 \end{pmatrix}.$$

Note that the curly bracket notation may only be used to populate arrays at the same time as when they are declared—for example the code

```
int array5[3] = {0, 1, 2};
```

is acceptable, but the code

```
int array6[3];
array6 = {0, 1, 2};
```

will not be accepted by the compiler.

1.4.6 ASCII Characters

ASCII characters are numbers, uppercase letters, lowercase letters and some other commonly used symbols: most of the characters on your keyboard are ASCII characters. Variables that are ASCII characters are declared using the keyword `char`. Example code using an ASCII character is shown below.

```
#include <iostream>

int main(int argc, char* argv[])
{
  char letter;
  letter = 'a'; // note the single quotation marks

  std::cout << "The character is " << letter << "\n";

  return 0;
}
```

1.4.7 Boolean Variables

Boolean variables take either the value `true` or the value `false`. These variables are commonly used when specifying whether a portion of code should be executed in conjunction with `if` and `while` statements (which will be introduced in Chap. 2). Examples of Boolean variables are given below.

```
1  bool flag1, flag2;
2  flag1 = true;
3  flag2 = false;
```

1.4.8 Strings

The data type `char` represents one ASCII character. A string may be thought of as an ordered collection of characters. For example, "C++" is a string consisting of the ordered list of characters "C", "+", and "+".

To use strings in C++ requires the header file `string`. The library which may be accessed using this header file contains significant functionality for the use and manipulation of strings. The bulk of coding for scientific computing applications requires operations on numerical variables, and so we do not discuss this data type in much detail. In the example code below, we demonstrate how to declare a string, how to determine the length of a string, how to access individual characters of the string, and how to print a string to the console.

A string in C++ is a little like an array of characters together with a layer of extra functionality. There is no need to understand *why* the length and elements of the string may be accessed in this way: an understanding of *how* is sufficient.

```
1   #include <iostream>
2   #include <string>
3
4   int main(int argc, char* argv[])
5   {
6      std::string city; // note the std::
7      city = "Oxford"; // note the double quotation marks
8      std::cout << "String length = " << city.length() << "\n";
9      std::cout << "Third character = " << city.at(2) << "\n";
10     std::cout << "Third character = " << city[2] << "\n";
11     std::cout << city << "\n"; // Prints the string in city
12     std::cout << city.c_str() << "\n"; // Also prints city
13  }
```

In line 9 and line 10 of the code recall that arrays in C++ have indices that begin from zero: `city.at(2)` and `city[2]` both refer to the entry of the array of characters with index 2, that is, "f", the third letter of the string "Oxford". Lines 11 and 12 both have the effect of printing the contents of `city` ("Oxford") to the screen. Line 12 prints the contents of `city` to the screen, but does so by first converting from a C++ string to a C string, which is an array of type `char`. The string utility function `c_str` is not needed here, but is useful in cases where we need to pass a C++ string to a function which expects an array of type `char`.

1.5 Simple Input and Output

It would be pointless to write a code without having the means to communicate the output of the code to the user, or to some other application. As such, *output* is a programming technique that must be mastered by all programmers. Similarly, the user of software would expect to be provided with the ability to specify data that the software would use to generate output: *input* is therefore just as important a programming skill. We now describe basic C++ commands to allow output to the screen and input from the keyboard. In Chap. 3, we provide a fuller explanation, describing input from, and output to, a file, and a more flexible specification of the format of this output.

1.5.1 Basic Console Output

We have already briefly discussed console—or screen—output in Sect. 1.2, and have seen that the statement

```
std::cout << "Hello World\n";
```

prints the text "Hello World" to the screen, followed by a new line.

We may use `std::cout` to write more than one entity to the console at a time. This is best explained by example: consider the statements below.

```
1   int x = 1, y = 2;
2   std::cout << "x = " << x << " and y = " << y << "\n";
```

The second statement above tells the computer to first print the string "x = ", followed by the value assigned to the variable x, then the string " and y = ", then the value assigned to the variable y, and finally to finish with a new line. The output is therefore

```
x = 1 and y = 2
```

Note that any spaces required in the output must be included within quotation marks in the statement that begins `std::cout`.

We have already seen one formatting command for output in C++: the new line formatting command `\n`. Some other useful formatting commands are shown in Table 1.2.

Table 1.2 Some formatting commands for console output

Command	Symbol
new line	\n
tab	\t
'	\'
"	\"
?	\?
bell sound	\a

Output from C++ is *buffered*. Sometimes, for example, if the computer is busy doing a large volume of computation, the program may not print the output to the screen immediately. If immediate output is desirable then use the statement "`std::cout.flush();`" after any `std::cout` command to ensure the output is printed before any other statements are executed, as shown in the listing below. As with certain aspects of string manipulation discussed in Sect. 1.4.8, at this stage it is sufficient to understand how to send output to the console immediately without worrying why it is done in this way.

```
1    std::cout << "Hello World\n";
2    std::cout.flush();
```

1.5.2 Keyboard Input

Keyboard input for numerical variables and characters is achieved using the input stream `std::cin`, where `cin` is a contraction of **console in**. As with console output, the `iostream` header file must be included. The following code prompts someone to enter their Personal Identification Number—commonly known as their PIN—and then assigns the number entered to the integer variable `pin`.

```
1    int pin;
2    std::cout << "Enter your PIN, then hit RETURN\n";
3    std::cin >> pin;
```

`std::cin` may be used to ask for more than one input at a time, as shown below.

```
1    int account_number, pin;
2    std::cout << "Enter your account number\n";
3    std::cout << "and then your PIN followed by RETURN\n";
4    std::cin >> account_number >> pin;
```

Keyboard input for variables of type string is slightly different. An example of how to input a string is given below. As with the commands for basic manipulation of strings given in Sect. 1.4.8, we do not attempt to explain why strings are input in this way: this will become clear when more advanced features of C++ are explained later in this book.

```
1    #include <iostream>
2    #include <string>
3
4    int main(int argc, char* argv[])
5    {
6      std::string name;
7      std::cout << "Enter your name and then hit RETURN\n";
8      std::getline(std::cin, name);
9      std::cout << "Your name is " << name << "\n";
10
11     return 0;
12   }
```

1.6 The assert Statement

Scientific computing applications usually require a massive number of complicated mathematical computations. If any one of these computations is incorrect, then the final results of the computation will usually be incorrect. Finding the source of the error is an excruciatingly tedious process, and so we strongly recommend the use of the features of the C++ language that allow identification of unexpected occurrences such as an attempt to compute the square root of a negative number.

In Chap. 9 we point to the notion that there are various levels or degrees of error. In particular, we introduce *exceptions*, which are a feature of the C++ language that allow very effective handling of an unexpected occurrence when a code is being run. A less sophisticated approach is to use assert statements, as demonstrated in the code below. Note the inclusion of the extra header file cassert that is required to use assert statements.

```cpp
1  #include <iostream>
2  #include <cassert>
3  #include <cmath>
4
5  int main(int argc, char* argv[])
6  {
7      double a;
8      std::cout << "Enter a non-negative number\n";
9      std::cin >> a;
10     assert(a >= 0.0);
11     std::cout << "The square root of "<< a;
12     std::cout << " is " << sqrt(a) << "\n";
13     return 0;
14 }
```

The code above invites the user to enter a nonnegative number, and returns the square root of this number. Before the square root is calculated, we check that the number really is nonnegative through the `assert` statement. We will see in Chap. 2 that the ">=" that appears in line 10 of the code is the "greater than or equal to" operator: this line of code therefore checks that the variable a is nonnegative. To see the effect of the `assert` statement, we first save the code as `program.cpp` and then compile the code without any optimisation flags to produce executable `a.out`. If, when this executable is run, the number −5 is entered, the code terminates at the `assert` statement with the following error message.

```
a.out:: program.cpp:10: int main(int, char**): Assertion 'a >= 0.0' failed
```

A further C++ function that is useful in conjunction with assertions is the function `std::isfinite`. This allows confirmation that a variable x contains a finite value, and not an infinite value (obtained, for example, by dividing a non-zero number by zero) or some other value that is not defined as a number (such as the square-root or logarithm of a negative number).[5] The use of this function along with an assert statement is illustrated in the code fragment below.

```cpp
double x;
assert(std::isfinite(x));
```

Although we emphasise that this is a very rudimentary technique for identifying errors, and that we will introduce more sophisticated techniques later, `assert` statements can provide significant information: in the error message above we see that the exact line of code where the problem occurred has been identified. Another

---

[5] For those values which fail the `std::isfinite` test it is possible to differentiate between infinite numbers (using `std::isinf`) and those which are "not a number" (using `std::isnan`).

advantage of `assert` statements is that they can be automatically removed when the code is compiled with the "`-DNDEBUG`" flag. This allows you to test code with the assertions activated but to distribute a faster program that has the assertions deactivated by compiling using the command

```
g++ -DNDEBUG program.cpp
```

1.7 Tips: Debugging Code

There are many tools designed to aid with the debugging of code. The most basic of these is the compiler, and the flags associated with the compiler, as described in Sects. 1.3.2 and 1.3.3. More sophisticated tools exist, but they are aimed at larger scale projects, such as those that we will develop in later chapters of this book.

Rather than learning to use a sophisticated debugging tool whilst in the early stages of learning C++, we suggest below some simpler techniques for debugging the code that you will be writing when tackling the exercises in the early chapters of this book.

Compile your code frequently.   Saving your code and compiling it using the warning compiler flag described in Sect. 1.3.3 every time a few statements are added is a useful diagnostic to see if any potential problems are being introduced. If there are any problems, comment out the new statements and recompile. Then add the statements in one at a time until the problem line is identified. When you first write code in C++ you may be amazed how often you forget the basic syntax such as a semi-colon at the end of a statement.

Save your project frequently.   If you have code that works and you need to add new functionality, then do not throw away the old version. If things go wrong then you will be able to see exactly what you changed and if all else fails you will have a working version to roll back to. If it is critical that you are able to roll back to a working version of the code, or if you are in a collaborative project, we recommend that you use a version control system.[6]

Always test the code with a simple example.   For example, if you are writing code to add the elements of two arrays verify the output by comparison with a calculation that you have carried out yourself.

Understand errors that arise when executing the code.   If your program complains of a "segmentation error" when executing, it is likely that you have attempted to access a member of an array that is out-of-range: that is, you may have attempted to access the 6th entry of an array that was only declared to have 4 elements.

---

[6]There are many open source version control systems such as CVS, Subversion, Mercurial or Git to help you with this. There are also organisations who will host your code repository for you.

Use output.   If you need to know where your program is crashing, and why, then print out some values of variables at key points in the execution. Do not forget to `flush` the output so that it appears before the program crashes!

Use assertions.   If you expect a certain property at the start of a section of code, for example, that the scale factor is nonzero or that the argument of a square-root is nonnegative, you can check for it using assertions (introduced in Sect. 1.6).

C++ arrays are indexed beginning from zero.   If the array `temperature` is declared as having 4 elements, the statement "`temperature[4] += 1.0;`" will cause problems.

Use a debugger.   If all else fails then debug your program using a debugger. Tips on using a debugger are to be found in Sect. 7.7.

1.8 Exercises

1.1 To ensure that your compiler is correctly set up, copy and save the file `Hello-World.cpp` displayed in Listing 1.1, compile it, and execute it.

1.2 Write code that asks a user to enter two integers from the keyboard and then writes the product of these integers to the screen.

1.3 Write code that declares two vectors as arrays of double precision floating point numbers of length 3 and assigns values to each of the entries. Extend this code so that it calculates the scalar (dot) product of these vectors and prints it to screen. Finally, extend the code so that it prints the Euclidean norm of both vectors to screen.
[*See Sect.* A.1.2 *for a definition of the scalar product, and Sect.* A.1.5 *for a definition of the Euclidean norm of a vector.*]

1.4 Write code that declares four 2×2 matrices of double precision floating point numbers, A, B, C, D, and assigns values to the entries of A and B. Let C = A + B, and D=A*B. Extend your code so that it calculates the entries of C and D, and then prints the entries of these matrices to screen.

1.5 Write code that invites the user to input separately strings that store their given name and their family name. Print the user's full name to screen.

1.6 I want to record the number of cars that drive past my house each day for five consecutive days, and calculate the average of these numbers. Create an integer array to store these five numbers, and then write code to calculate the average of these numbers. Execute your code using the sample data 34, 58, 57, 32, 43. Verify that you get the correct answer of 44.8.

[*Hint: read the material in Sect.* 1.4.4 *on converting integers to double precision floating point numbers.*]

1.7 Investigate the use of the compiler error warning flags discussed in Sect. 1.3.3. For example: (i) declare an integer as a constant variable and then attempt to change this value later in the code; and (ii) attempt to set an integer variable to the value 3.2.

Flow of Control

In almost any computer program written for a scientific computing application, we need to allow the computer to execute a collection of statements if—and only if—some criterion is met. For example, if we were writing a program to control the motion of a spacecraft travelling to Mars, the program would include lines of code that would control the safe landing of the spacecraft. As the craft completes its touchdown, it fires retrorocket motors to control descent until the sensors detect that the landing gear is in contact with the planet's surface. It is imperative that the lines of code which say "cut the motor if and only if there is a strong signal from the landing gear" are executed at exactly the right time. If these instructions are not executed when the spacecraft has landed, the retrorockets may fire for too long and cause damage to the craft. On the other hand, if the instruction to cut the motors is executed when the spacecraft is still descending, we would expect the spacecraft to crash.[1] It is clear that the relevant lines of code should be executed if, and only if, certain conditions are met.

As with most programming languages, conditional branching may be achieved in C++ programs by using an `if` statement. Similarly, we may use a `while` statement to execute a collection of statements until a specified condition is met, and a `for` loop to execute a collection of statements a specified number of times. In this chapter, we explain how to utilise these features of the C++ language.

---

[1]Nobody knows what happened to the Mars Polar Lander in the last few seconds of its descent in 1999, but experts believe there was a bug in the landing gear sensor code. This bug involved accumulating weak signals from the landing gear and may have caused the retrorockets to cut out too early.

© Springer International Publishing AG, part of Springer Nature 2017
J. Pitt-Francis and J. Whiteley, *Guide to Scientific Computing in C++*, Undergraduate Topics in Computer Science,
https://doi.org/10.1007/978-3-319-73132-2_2

2.1 The if Statement

The most basic use of an if statement is to execute one or more statements if, and only if, a given condition is met. As we shall see in this section, we may build upon this simple construct to write more complicated statements when required.

2.1.1 A Single if Statement

Let us suppose that we wish to execute two statements, Statement1 and Statement2, if—and only if—the condition p > q is met. The following code demonstrates the basic syntax for this in C++.

```
1    if (p > q)
2    {
3      Statement1;
4      Statement2;
5    }
```

If the condition p > q is met, then the code enclosed by the curly brackets is executed. The condition (in round brackets) is technically know as the *guard*. Note the indentation within the curly brackets in the above listing. While this is not necessary for the compiler to understand the meaning, it makes it clearer to the reader which statements are executed if the condition p > q is met.

If only one statement—Statement1—is to be executed when the condition p > q is satisfied, then curly brackets are not strictly necessary. For example, the following two code fragments will execute Statement1 if the condition p > q is met.

```
1    if (p > q)
2      Statement1;
```

or

```
     if (p > q) Statement1;
```

Although either of these two variants of the code will do what we want it to, we do not recommend them, as the curly brackets make it very clear precisely which statements are executed as a consequence of a given if statement. As such, we would strongly suggest the use of curly brackets, as shown in the code below. More suggestions on tips for ensuring code is clearly readable—known as *coding conventions*—may be found in Sect. 6.6.

```
1   if (p > q)
2   {
3      Statement1;
4   }
```

2.1.2 Example: Code for a Single if Statement

Below is a concrete example of code that uses an if statement. This code changes the value of x to zero if, and only if, x is negative. If x is not negative, line 5 of the code will not be executed, and the value of x will be unchanged.

```
1   double x = -2.0;
2
3   if (x < 0.0)
4   {
5      x = 0.0;
6   }
```

2.1.3 if-else Statements

It is often the case that we want to set a variable to one value if a specified condition is met, and to a different value otherwise. This may be implemented in C++ code by the use of an if statement in conjunction with an else statement. The fragment of code below sets the double precision floating point variable y to the value 2 if the integer variable i is positive, and to the value 10 otherwise.

```
1    int i;
2    //...
3    double y;
4    if (i > 0)
5    {
6       y = 2.0;
7    }
8    else
9    {
10      //When i <= 0
11      y = 10.0;
12   }
```

Note the comment in line 10 of the listing above. As no condition is needed for the else condition, it is always good programming practice to use a comment to explicitly state under what conditions the else condition should be met.

2.1.4 Multiple `if` Statements

We may extend the `if` and `else` statements described above to allow more complicated conditions on the execution of statements. Extending the previous example, suppose the double precision floating point variable y takes the value 2 if the integer variable i is greater than 100, y takes the value 10 if i is negative, and y takes the value 5 otherwise. C++ code for this condition is given below.

```
1   int i;
2   //...
3   double y;
4   if (i > 100)
5   {
6      y = 2.0;
7   }
8   else if (i < 0)
9   {
10     y = 10.0;
11  }
12  else
13  {
14     //When 0 <= i <= 100
15     y = 5.0;
16  }
```

2.1.5 Nested `if` Statements

It is common in scientific computing to have an algorithm where statements must be executed if, and only if, two separate conditions are met. One way of implementing this is to use *nested* `if` statements, as shown below. In this code the double precision floating point variable y is assigned the value 10 if, and only if, the conditions $x > z$ and $p > q$ are both met.

Listing 2.1 A nested `if` statement

```
1   double x, z, p, q;
2   double y;
3   if (x > z)
4   {
5      if (p > q)
6      {
7         //Both conditions have been met
8         y = 10.0;
9      }
10  }
```

2.1.6 Boolean Variables

Boolean variables may be used as the condition with an if statement. This is demonstrated in the fragment of code below.

```
1   bool flag = true;
2   if (flag)
3   {
4      std::cout << "This will be printed\n";
5   }
6   else
7   {
8      // flag is false
9      std::cout << "This won't be printed\n";
10  }
```

2.2 Logical and Relational Operators

In Sect. 2.1 we demonstrated the use of if statements by using the relational operator "greater than". To fully utilise if statements and, as we shall see later, while statements and for loops, we need to extend our range of logical and relational operators. These are summarised in Tables 2.1 and 2.2. The combination of logical and relational operators allow any reasonable condition to be implemented in C++ code.

A first example of the combination of logical and relational operators is to replace the nested if statements in Listing 2.1 by a single if statement. The condition in the new if statement is true if, and only if, both the condition $x > z$ and the condition

Table 2.1 Logical operators in C++

Logical condition	Operator
AND	&&
OR	||
NOT	!

Table 2.2 Relational operators in C++

Relation	Operator
Equal to	== (note that it is not "=")
Not equal to	!=
Greater than	>
Less than	<
Greater than or equal to	>=
Less than or equal to	<=

p > q are true. If this compound condition is met, the value 10 is assigned to the variable y. This is demonstrated in the code below.

```
1    double x, z, p, q;
2    double y;
3    if ((x > z) && (p > q))
4    {
5        //Both conditions have been met
6        y = 10.0;
7    }
```

The example code fragment below uses a combination of logical and relational operators to set a double precision floating point variable y to the value 10 if either p > q or the integer variable i is not equal to 1. If neither of these conditions has been met, then the variable y is assigned the value −10.

```
1    double p, q;
2    int i;
3    double y;
4    if ((p > q) || (i != 1))
5    {
6        //One or both conditions have been met
7        y = 10.0;
8    }
9    else
10   {
11       //Neither condition has been met: p<=q and i==1
12       y = -10.0;
13   }
```

The logical operator "NOT" is often used in conjunction with Boolean variables. This is demonstrated in the example code below, where the integer variable i is incremented by the value 2 if, and only if, the Boolean variable flag takes the value false.

```
1    int i;
2    bool flag = false;
3    if (!flag)
4    {
5        // !flag is true when flag is false
6        i += 2;
7    }
```

2.3   The while Statement

A while statement is used if a collection of statements are to be executed until some prescribed condition is not met. The C++ syntax for while statements is similar to that for if statements.

A first example of a while statement is given below. A variable x is initially assigned the value 10. On each execution of the code inside the while statement the value of the variable x is halved. This is repeated while the value of the variable x is greater than 1.

Listing 2.2   A while loop

```
1  double x = 10.0;
2  while (x > 1.0)
3  {
4      // This loop will execute while x > 1, so if the
5      // value of x does not decrease then it will not
6      // terminate.
7      x *= 0.5;
8  }
9  // Here we know the guard (x > 1.0) has broken.
10 // This means that after the loop, x <= 1.0
```

Although while statements are frequently used in C++ programming, they should be used with care. Consider the fragment of code below. Suppose we want to develop Listing 2.2 above so that we count the number of times that we halve the variable x. This may be achieved by the use of an integer variable count which is incremented every time the statements inside the curly brackets are executed, as shown below.

```
1  double x = 10.0;
2  int count = 0;
3  while (x > 1.0)
4  {
5      x *= 0.5;
6      std::cout << "x = " << x << ", count = "
7                << count << "\n";
8      count++;
9      std::cout << "x = " << x << ", count = "
10               << count << "\n";
11     std::cout << "Reached bottom of while loop\n";
12 }
13 std::cout << "count = " << count << "\n";
```

The output of this code is shown below.

```
x = 5, count = 0
x = 5, count = 1
Reached bottom of while loop
x = 2.5, count = 1
x = 2.5, count = 2
Reached bottom of while loop
x = 1.25, count = 2
x = 1.25, count = 3
Reached bottom of while loop
x = 0.625, count = 3
x = 0.625, count = 4
Reached bottom of while loop
count = 4
```

The important thing to note in the example output above is that the condition $x > 1.0$ is tested *only at the beginning* of the statements enclosed within the curly brackets. In particular, this condition first became untrue when the variable x was assigned the value 0.625 at line 5 in the code. However, the condition $x > 1.0$ was not tested at this point, and so the variable count was incremented as line 8 will be executed before leaving the while loop.

Were we to want a loop to be executed *at least once*, regardless of any other conditions, then we can use the do-while syntax which tests at the end of the loop, as shown below.

```
1   double x = 0.8;
2   int count = 0;
3   do
4   {
5     x *= 0.5;
6     std::cout << "x = " << x << ", count = "
7                << count << "\n";
8     count++;
9     std::cout << "x = " << x << ", count = "
10               << count << "\n";
11    std::cout << "Reached bottom of do-while loop\n";
12  } while (x > 1.0);
13  std::cout << "count = " << count << "\n";
```

The output of this code (shown below) demonstrates that the body of the loop is executed once, even though the initial value of x does not satisfy the condition in the guard.

```
x = 0.4, count = 0
x = 0.4, count = 1
Reached bottom of do-while loop
count = 1
```

We may nest while statements in exactly the same way as if statements, described in Sect. 2.1.5.

2.4 Loops Using the for Statement

The simplest application of a for loop is to execute a collection of statements a specified number of times. The fragment of code below demonstrates how to execute a given statement 10 times.

```
1   for (int i=0; i<10; i++)
2   {
3     std::cout << i << " ";
4   }
```

Line 1 of the code above deserves more explanation. The first statement in this line of code declares an integer variable i, and initialises this variable to the value 0. The code inside the curly brackets is executed if, and only if, the variable i is less than 10. The final content of this line of code increments i by the value 1 each time all the statements enclosed by the curly brackets have been executed. The output of this code is therefore

```
0 1 2 3 4 5 6 7 8 9
```

We may also nest for loops in a similar way to that for if statements described in Sect. 2.1.5. Furthermore, for loops may be defined to be executed a variable number of times, as demonstrated in the example code below.

```
1   for (int i=0; i<5; i++)
2   {
3     for (int j=5; j>i; j--)
4     {
5       std::cout << "i = " << i
6                 << "  j = " << j << "\n";
7     }
8   }
```

Before explaining what the code above does, it is important to understand what line 3 of code (the second `for` statement) does. In a similar vein to the discussion of the initial example of a `for` loop, we see that the first statement initialises the integer variable `j` to 5. The statements within the furthest indented curly brackets are executed when the variable `j` is greater than the variable `i`. Each time these statements have been executed, `j` is decremented by the value 1.

We are now in a position to understand the whole of the code above. The loop over the variable `i` is known as the *outer loop*, and the loop over the variable `j` is known as the *inner loop*. The first time the statements in the outer loop are executed, `i` takes the value 0. When `i` takes this value, the third line of code tells us that `j` takes the values 5, 4, 3, 2, 1. The second time the statements in the outer loop are executed, `i` will take the value 1, and so `j` will take the values 5, 4, 3, 2. We may now deduce that the output of the code above will be

```
i = 0    j = 5
i = 0    j = 4
i = 0    j = 3
i = 0    j = 2
i = 0    j = 1
i = 1    j = 5
i = 1    j = 4
i = 1    j = 3
i = 1    j = 2
i = 2    j = 5
i = 2    j = 4
i = 2    j = 3
i = 3    j = 5
i = 3    j = 4
i = 4    j = 5
```

2.4.1 Example: Calculating the Scalar Product of Two Vectors

The scalar product between two vectors of the same length may be computed using a `for` loop. Suppose the vectors are both of length n, and are stored in double precision floating point arrays `vector1` and `vector2` of the correct size. Remembering that the indexing of C++ arrays begins from zero, the scalar product (discussed in more detail in Sect. A.1.2) between these vectors—defined to be a double precision floating point variable `scalar_product`—is given mathematically by the following sum:

$$\texttt{scalar\_product} = \sum_{i=0}^{n-1} \texttt{vector1[i]} \times \texttt{vector2[i]}.$$

The mathematical expression above for calculating the scalar product is implemented in C++ below for the case n=2. Note that the variable `scalar_product` must be initialised to 0 before any calculation is carried out.

```
1    double vector1[2], vector2[2];
2    vector1[0] = 0.5; vector1[1] = -2.3;
3    vector2[0] = 34.2; vector2[1] = 0.015;
4    double scalar_product = 0.0;
5    for (int i=0; i<2; i++)
6    {
7        scalar_product += vector1[i] * vector2[i];
8    }
```

2.5 The switch Statement

A good understanding of the flow of control resulting from if, while and for statements is crucial for implementation of scientific computing applications. One further statement that is used less frequently is the switch statement. This statement is best explained by example. Consider the code below, where the variable i has been declared as an integer. Note that the language specification says that the *control variable*, which is i in our case, must be an integer and not a floating point type.

```
1    int i;
2    switch(i)
3    {
4        case 1:
5            std::cout << "i = 1\n";
6        case 20:
7            std::cout << "i = 1 or i = 20\n";
8            break;
9        default:
10           std::cout << "i is not 1 or 20\n";
11   }
```

If i takes the value 1 when the code above is executed, the statements below line 4 will be executed until the line of code break is reached (line 8). At the point when break is reached, the flow of execution will leave the code inside the curly brackets. Similarly, if the code is executed when i takes the value 20, then the statements below line 6 will be executed until the line of code break is reached. For all other values of i the line of code after default (line 9) will be executed.

Switch statements were introduced to programming languages because they are very easy for compilers to implement efficiently. However, they are notorious as places where programmers introduce bugs by forgetting to end case statements with the break keyword or by forgetting to give a default case. Switch statements should be written with care.

2.6 Tips: Loops and Branches

In this tips section, we highlight several traps that programmers who are new to C++ may fall into.

2.6.1 Tip 1: A Common Novice Coding Error

Below is code that has been written with the intention of doubling a variable x five times.

```
1   double x = 2.0;
2   for (int i=0; i<5; i++);
3   {
4     x *= 2.0;
5   }
6   std::cout << "x = " << x << "\n";
```

It would be expected that this code would output the value $2 \times 2^5 = 64$. However, the actual output of this code is

```
x = 4
```

Why is this? Hint: look very closely at line 2 of the code above.

The reason for the surprising output is the semi-colon at the end of line 2. This is a common error for programmers who are new to the language. After seeing that most lines end with a semi-colon you might begin to get into the habit of ending *every* line with one. When you see the guard at the beginning of a for, while or if statement without a semi-colon at the end then it might be tempting to stick one in!

You might ask "If the loop is not executing as intended, why is the final answer x = 4 and not x = 2?". The answer is that the empty space in line 2 between the ")" and the ";" is being interpreted as the body of the loop—it is the empty *nothing* which is executed 5 times. The intended body of the loop (lines 3–5) is treated as a *block* with special scope (see Sect. 5.1 for more information). This block has no connection with the for loop and is executed once.

2.6.2 Tip 2: Counting from Zero

Programmers who are experienced with MATLAB or Fortran may be used to a loop beginning from 1 and ending when the loop variable reaches a given value. If we wish a loop to execute exactly four times, we would write it in MATLAB or Fortran as

```
1  %MATLAB loop
2  for j=1:4,
3      j
4  end
```

```
1  ! Fortran loop
2     DO 10 J = 1, 4
3        WRITE(*,*) 'J = ', J
4  10 CONTINUE
```

In both cases the variable j (in the MATLAB code) or J (in the Fortran code) takes values from 1 to 4 inclusive. When programming in C++ it is common to write the equivalent loop from 0 up to, but not including, 4. That is, $j = 0, 1, 2, 3$. The reason for this is that while MATLAB and Fortran use *one-based indexing* where array indexing starts at 1, C++ uses *zero-based indexing*. It is a good idea to write loops in the form of the second loop given below.

```
1   // This loop is natural for MATLAB programmers
2   for (int j=1; j<=4; j++)
3   {
4      std::cout << "j = " << j << "\n";
5   }
6   // This loop is natural for C++ programmers
7   for (int j=0; j<4; j++)
8   {
9      std::cout << "j = " << j << "\n";
10  }
```

2.6.3   Tip 3: Equality Versus Assignment

When we introduced relational operators in Table 2.2, we noted that there is a difference between a single = and a double ==. The operator = is an assignment operator which takes the value on the right-hand side and assigns it to the variable on the left-hand side. The equality operator == returns true if, and only if, the values on the left and right are equal.

A common programming error is to mistake one for the other.

```
1   // This erroneous line has no effect
2   x == 2+2;
3   // After testing x against the value 4, the true/false
4   // answer is discarded.
5
6   x = 3;
7   //This erroneous line will alter the value of x
8   if (x = 4)
9   {
10     x = 6;
11  }
```

The code above shows two common unintended bugs in C++ code. Line 2 of this code will test whether or not the variable x is equal to 4, but assign no value to x. This line therefore has no overall effect. Your compiler may give you a warning. However, as different compilers will give different warnings, you should not rely on this. Unless suitable compiler flags are used the compiler will give no error since it is valid syntax. The second error is shown in lines 8–11 of the code. In this case, line 8 of the code uses assignment (a single equals sign) when equality testing (a double equals sign) was intended. This code will have the effect of changing the value of x to the value 4 when this was not intended. The condition which is actually tested is obtained from the value of the assignment. A non-zero value (in this case the value 4) is interpreted as success, and so this condition is met. The code inside the curly brackets therefore will be executed, and so the variable x will take the value 6. Again, this is valid syntax so the compiler may give no warning or error.

Some compilers may report these types of problems as either warnings or errors. You may be able to ensure that the compiler informs you of these quite subtle problems by switching on warnings, as we described in Sect. 1.3.3.

If we include the above in a program called Tip.cpp, and compile with the flag to switch on all warnings, then the GNU C++ compiler gives the following warnings:

```
$ g++ -Wall Tip.cpp
Tip.cpp: In function 'int main(int, char**)':
Tip.cpp:2: warning: statement has no effect
Tip.cpp:8: warning: suggest parentheses around assignment
used as truth value
```

We see that, although the offending lines are not doing what was intended, an executable that can be run is still produced. If we compile with the compilation flag –Werror discussed in Sect. 1.3.3, then the warnings now become errors, and so no executable program is produced. In this case, we get the following output at compilation time:

```
$ g++ -Wall -Werror Tip.cpp
cc1plus: warnings being treated as errors
Tip.cpp: In function 'int main(int, char**)':
Tip.cpp:2: error: statement has no effect
Tip.cpp:8: error: suggest parentheses around assignment used
as truth value
```

2.6.4 Tip 4: Never Ending while Loops

As discussed briefly in Sect. 2.3, it is essential to ensure that the code can always leave a while loop. The code below was written to find the maximum of an array of four positive numbers called positive_numbers. Why will this code never leave the while loop?

```
 1   double positive_numbers[4] = {1.0, 5.65, 42.0, 0.01};
 2   double max = 0.0;
 3   int count = 0;
 4   while (count < 4)
 5   {
 6     if (positive_numbers[count] > max)
 7     {
 8       max = positive_numbers[count];
 9     }
10   }
```

The problem with the code above is that the integer count is not incremented inside the while statement. The variable count will therefore always take the value 0, the condition count < 4 will always be satisfied, and the code will never exit the while loop.

2.6.5 Tip 5: Comparing Two Floating Point Numbers

If i and j have been declared as integers, and we want to set another integer variable k to zero if these variables take the same value, then this may easily be written in C++ using the following code.

```
 1   int i, j, k;
 2   if (i == j)
 3   {
 4     k = 0;
 5   }
```

Suppose, instead, we wanted to set k to zero if two double precision floating point variables p and q take the same value. It may be thought that a very simple modification of the code above will suffice, where p and q are declared as double precision floating point variables and the guard in line 2 of the listing is modified to test for equality of p and q. This, however, is not the case. Operations between floating point numbers all induce rounding errors. As a consequence, if the true value of a calculation is 5, the number stored may be 5.000000000000186. Testing two double precision floating point variables for equality is unlikely to give the expected answer, as due to rounding errors it is unlikely that two such variables will ever be equal. Instead, we should check that the two numbers differ by less than some very small number,[2] as shown below.

```
1    double p, q;
2    int k;
3    if (fabs(p-q) < 1.0e-8)
4    {
5       k = 0;
6    }
```

2.7 Exercises

2.1 Below is an example fragment of code that uses several features introduced in this chapter. The variables x, y and z are all double precision floating point variables.

```
1    double x, y, z;
2    if ((x > y) || (x < 5.0))
3    {
4       z = 4.0;
5    }
6    else
7    {
8       z = 2.0;
9    }
```

1. Explain, in words, what the fragment of code does.
2. What value would the fragment of code assign to the variable z when the variables x and y take the following values:

---

[2]If p and q are the results of two calculations which ought to be equal, to within machine precision, then they many differ by about $|p| \times$ DBL_EPSILON, since DBL_EPSILON $\sim$2e–16 is defined in #include <cfloat> to be smallest double precision floating point number such that 1.0+DBL_EPSILON is not equal to 1.0 when rounding errors are taken account of.

(a) $x = 10.0$, and $y = -1.0$;
(b) $x = 10.0$, and $y = 20.0$; and
(c) $x = 0.0$, and $y = 20.0$.

3. Modify the code above so that the condition $x > y$ is replaced by $x \geq y$.

2.2 Below is some example code. The exercises below all require modification of this code. In all cases use a suitable check to ensure your code is correct.

```cpp
1  #include <iostream>
2
3  int main(int argc, char* argv[])
4  {
5      double p, q, x, y;
6      int j;
7
8      return 0;
9  }
```

1. Set the variable x to the value 5 if either p is greater than or equal to q, or the variable j is not equal to 10.
2. Set the variable x to the value 5 if both y is greater than or equal to q, and the variable j is equal to 20. If this compound condition is not met, set x to take the same value as p.
3. Set the variable x according to the following rule.

$$x = \begin{cases} 0, & p > q, \\ 1, & p \leq q, \text{ and } j = 10, \\ 2, & \text{otherwise.} \end{cases}$$

2.3 In this exercise you are asked to write and test a program which sums a list of numbers which are provided by a user via $std::cin$ (see Sect. 1.5.2).

1. Write a program that calculates the sum of a collection of positive integers that are entered by the user from the keyboard. Your program should prompt the user to enter each integer followed by the return key, and to enter "−1" at the end of the list of integers to be added. Note that there is no need to store the list of integers: you can keep track of the sum as the user is entering the values.
2. Modify your code so that the code terminates if the sum of integers entered up to that point exceeds 100.
3. Modify your code so that, if the user has entered an incorrect integer, they may enter "−2" to reset the sum to zero and begin entering integers again.

2.4 This exercise uses the following vectors and matrices:

$$u = \begin{pmatrix} 1 \\ 2 \\ 3 \end{pmatrix} ; v = \begin{pmatrix} 6 \\ 5 \\ 4 \end{pmatrix} ; A = \begin{pmatrix} 1 & 5 & 0 \\ 7 & 1 & 2 \\ 0 & 0 & 1 \end{pmatrix} ; B = \begin{pmatrix} -2 & 0 & 1 \\ 1 & 0 & 0 \\ 4 & 1 & 0 \end{pmatrix} .$$

Furthermore, the vector w satisfies $w = u - 3v$. These vectors and matrices are stored in arrays using the following program. This program includes code to calculate the vector w.

```cpp
#include <iostream>

int main(int argc, char* argv[])
{
   double u[3] = {1.0, 2.0, 3.0};
   double v[3] = {6.0, 5.0, 4.0};
   double A[3][3] = {{1.0, 5.0, 0.0},
                     {7.0, 1.0, 2.0},
                     {0.0, 0.0, 1.0}};
   double B[3][3] = {{-2.0, 0.0, 1.0},
                     {1.0, 0.0, 0.0},
                     {4.0, 1.0, 0.0}};

   double w[3];
   for (int i=0; i<3; i++)
   {
      w[i] = u[i] - 3.0*v[i];
   }

   return 0;
}
```

We now define vectors x, y, and z, and matrices C and D, such that

$$x = u - v,$$
$$y = Au,$$
$$z = Au - v,$$
$$C = 4A - 3B,$$
$$D = AB.$$

Develop the program above to calculate the vectors x, y, and z and the matrices C and D, using loops where possible. Hint: make sure you define arrays of an appropriate size for these variables. Check your answer by printing out the results, and comparing with direct calculation.

2.5 The inverse of a 2×2 square matrix is given in Sect. A.1.3.

1. Write code to calculate the inverse of the matrix given by

$$A = \begin{pmatrix} 4 & 10 \\ 1 & 1 \end{pmatrix}.$$

2. Check that the inverse calculated is correct by printing out the entries of the inverse, and comparing with direct calculation.
3. Modify your code to include an `assert` statement that checks that the determinant of the matrix is nonzero.

2.6 The Newton–Raphson method (see, for example, Kreyszig [2]) is often used to solve nonlinear equations of the form $f(x) = 0$. This is an iterative algorithm: given an initial guess x_0, successive iterates satisfy

$$x_i = x_{i-1} - \frac{f(x_{i-1})}{f'(x_{i-1})}, \quad i = 1, 2, 3, \ldots.$$

This algorithm may be terminated when $|x_i - x_{i-1}| < \varepsilon$ for some user-prescribed ε.

In this exercise, we will apply the Newton–Raphson algorithm to the function $f(x) = e^x + x^3 - 5$, with initial guess $x_0 = 0$.

1. Write down (on paper) the Newton–Raphson iteration for this choice of $f(x)$.
2. By using a `for` loop, and an array for the iterates x_i, write a program that implements the Newton–Raphson iteration for $i = 1,2,3,\ldots,100$. Print out the value of x_i on each iteration, and confirm that the iteration does converge as i increases. At this stage, do not worry about terminating the iteration when ε is sufficiently small.
3. Think of a check that can be performed on the iterates x_i, as i becomes larger, that allows you to have confidence that your solution is correct. Implement this check in your program.
4. It is not necessary to store the value of x_i on each iteration to implement the Newton–Raphson algorithm. All that is needed is the previous iterate, x_{i-1}, and the current iterate, x_i. Modify your code so that the array representing x_i, $i = 1, 2, \ldots, 100$ is replaced by two scalar variables, `x_prev` and `x_next`.
5. Modify your code so that, by use of a `while` statement, the iteration terminates when |x_next-x_prev|$< \varepsilon$. Investigate the use of different values of ε.

File Input and Output

Being able to transfer data between applications is an essential requirement of most scientific computing software. For example, data defining the boundary of an object may be generated from an image processing application. This data may subsequently be used by many applications written by a variety of users. To allow exchange of data between applications in this manner requires us to store data in a clearly specified format. Reading and writing files to a given specification therefore plays a key role in scientific computing applications, and is the subject of this chapter.

3.1 Redirecting Console Output to File

We introduced basic C++ commands for writing text and the contents stored by a variable to the console in Sect. 1.5. On a Linux system this output may very easily be redirected to a single file rather than the screen. Should the executable file be called `SampleCode`, this output may be printed to the file `SampleOutput.txt` by executing at the command line, as described in Sect. 1.3.2, with the executable name being followed by a specification of the file to be written to, as shown below:

```
$ ./SampleCode > SampleOutput.txt
$
```

When output has been redirected to file in this way, you may prefer to print to screen errors encountered by the program. This can be done using `std::cerr` as shown below. The word `cerr` is a contraction of **console err**or.

© Springer International Publishing AG, part of Springer Nature 2017
J. Pitt-Francis and J. Whiteley, *Guide to Scientific Computing in C++*, Undergraduate Topics in Computer Science,
https://doi.org/10.1007/978-3-319-73132-2_3

```
1   int x, y;
2   if (y == 0)
3   {
4      std::cerr << "Error - division by zero\n";
5   }
6   else
7   {
8      // y not zero
9      std::cout << x/y << "\n";
10  }
```

The syntax for `std::cerr` is identical to that for `std::cout`. When the console output is not redirected to file there is no difference between the effect of these two commands. However, when output is redirected to a specified file, only the `std::cout` statements are redirected: the output from a `std::cerr` statement will still be printed to the screen. Should output from the code above be redirected to file, then the value given by dividing x by y will be written to the specified file unless the variable y takes the value zero. Under these circumstances, the message "`Error - division by zero`" will be printed to the screen instead.

3.2 Writing to File

In the previous section, we explained how all the output of an application may be printed to a single file. This may be adequate for some applications, but is definitely not adequate for all applications. For example, were we to write a code to calculate the finite element solution of a given differential equation we may want to store the nodes of the mesh in one file, the connectivity array defining the elements in another file, the finite element solution in another file, and—perhaps—the nodes comprising the individual faces of the elements in another file. We therefore need to be able to write output to more than one file. Although C++ offers an extremely large number of commands for printing to file, almost all file formats can be achieved by using a very small subset of these commands.

Writing to, or reading from, file requires the additional header file `fstream`. In the code below, we show how to write to file. We first declare an *output stream variable* `write_output` by specifying it as being of type `std::ofstream`, and also specify the filename "`Output.dat`" as shown in line 9. Line 10 then checks that the file has been successfully opened: we return to this point below. Writing to file is similar to console output, but replacing `std::cout` with `write_output` in line 13: this writes the entries of the arrays x and y to the file associated with the output stream variable, in this case `Output.dat`. Finally, in line 15, when all required data has been written to file, we "close the file handle". In Sect. 1.5.1, we explained that console output is buffered, and so the output may not immediately be written to the console. Output to file is also buffered: closing the file handle *flushes* the buffer: that is, all data that has been buffered is written to file before the computer executes any further statements. It is important that this is done: if another part of the program reads a file which is still being written to, then we cannot be certain what data—if any—has yet been written to disk. Closing the file handle has the further effect that no more data can be written to this file: this prevents the file being corrupted by mistakenly attempting to write further data. We note at this point that explicitly closing the file handle on line 15, and in many of our later examples, is actually redundant for the simple reason that the call to `close()` will be run automatically as the file handle is tidied when the `main` function finishes. However it is good practice for the novice programmer to make this call explicitly and thereby to know when to expect output from their program to be written to file.

Listing 3.1 Basic writing to file

```
 1  #include <cassert>
 2  #include <iostream>
 3  #include <fstream>
 4
 5  int main(int argc, char* argv[])
 6  {
 7     double x[3] = {0.0, 1.0, 0.0};
 8     double y[3] = {0.0, 0.0, 1.0};
 9     std::ofstream write_output("Output.dat");
10     assert(write_output.is_open());
11     for (int i=0; i<3; i++)
12     {
13       write_output << x[i] << " " << y[i] << "\n";
14     }
15     write_output.close();
16     return 0;
17  }
```

It is also possible to flush a buffer without closing the file handle. This is done in a similar way as for console output in Sect. 1.5.1, and is demonstrated below for the output stream variable `write_output`.

```
    write_output.flush();
```

We explained above that it is important to check that a file has been opened (line 10 of the Listing 3.1) before attempting to write any data to it. If the file cannot be opened—perhaps we did not have permission to write to that file, or a directory we have specified does not exist—then writing to the `ofstream` may cause no error even though writing to the file is not possible. For example, if in line 9 we renamed the location of the output file to a folder we are restricted from writing to as follows:

```
9   std::ofstream write_output("/etc/Output.dat");
```

then we might expect the program to fail as we are unlikely to have permission to
write to the folder /etc/. However, without the test for the file being open the code
will exit normally, producing no output file. This would clearly be very frustrating
for the user of the code.

The executable created from Listing 3.1 will create a new file, Output.dat, if
this file does not already exist. If this file does exist, the executable generated from
the listing above will delete the original file and write a new file with the same name:
the original contents of the file will be lost.[1] Whether or not the file Output.dat
existed before the code above was executed, after execution there will be a file called
Output.dat that is listed below.

Listing 3.2 The file Output.dat

```
0   0
1   0
0   1
```

The code in Listing 3.1 may do what was required, but it may not. Suppose that,
rather than deleting the file if it exists, we want our code to append data to the end
of this file. This would be achieved by modifying line 9 of Listing 3.1 to

```
9   std::ofstream write_output("Output.dat", std::ios::app);
```

If the file Output.dat did not exist and we were to execute the code in
Listing 3.1, with line 9 modified as shown above, we would then create the file
Output.dat shown in Listing 3.2. If we were then to execute the code a second
time we would then end up with the file Output.dat being modified as shown in
Listing 3.3 below.

Listing 3.3 Modified file Output.dat

```
0   0
1   0
0   1
0   0
1   0
0   1
```

---

[1]If you want to check for the existence of a file before opening an output stream to it then a simple
thing to do is to first attempt to read from it. See Exercise 3.1

3.2.1   Setting the Precision of the Output

The key formatting command for scientific computing applications is specification of
the precision of the output. This is demonstrated in the listing below. The number in
brackets after the `precision` commands specifies the number of significant figures
that the output is correct to. Note that when the precision is set to 10 significant figures
in line 15 of the listing below only eight significant figures will be printed: this is
because the variable x is only given to eight significant figures, and so the remaining
accuracy requested is redundant.

```cpp
1  #include <iostream>
2  #include <fstream>
3
4  int main(int argc, char* argv[])
5  {
6     double x = 1.8364238;
7     std::ofstream write_output("Output.dat");
8
9     write_output.precision(3); // 3 sig figs
10    write_output << x << "\n";
11
12    write_output.precision(5); // 5 sig figs
13    write_output << x << "\n";
14
15    write_output.precision(10); // 10 sig figs
16    write_output << x << "\n";
17    write_output.close();
18
19    return 0;
20 }
```

3.3   Reading from File

When reading from file we first need to declare an *input stream variable* in a similar
way to the output stream variable described in Sect. 3.2, and then specify the file that
we wish to read. As with output to file, the header file `fstream` should be included.
Reading the file is then performed in a similar way to that described for keyboard
input in Sect. 1.5.2, with `std::cin` replaced by the input stream variable. Suppose
we want to input the file `Output.dat` shown in Listing 3.3. We know that this file
has six rows and two columns, and so we may read this file using the code shown
in Listing 3.4. The assertion in line 9 ensures that `Output.dat` is on disk in the
correct location and with the correct access privileges: if not, the assertion is tripped
and the code is terminated.

Listing 3.4 Reading column formatted data

```
 1  #include <cassert>
 2  #include <iostream>
 3  #include <fstream>
 4
 5  int main(int argc, char* argv[])
 6  {
 7    double x[6], y[6];
 8    std::ifstream read_file("Output.dat");
 9    assert(read_file.is_open());
10    for (int i=0; i<6; i++)
11    {
12      read_file >> x[i] >> y[i];
13    }
14    read_file.close();
15    return 0;
16  }
```

In the code above, we knew that the file we were reading had six rows and two columns, and so we knew when writing this code that the statements inside the `for` loop had to be executed six times. In many scientific computing applications we will want to read a file, but do not know the length of the file in advance. For example, we may know that a file contains a list of the coordinates of an unknown number of points in two dimensions: the file therefore has two columns, but an unknown number of rows. We cannot use a `for` loop as we do not know how many times the statements in this loop need to be executed. Instead, we use the Boolean variable associated with the input stream variable `read_file.eof()`. This variable takes the value `true` when the end of the file is reached, and allows us—through the use of a `while` statement—to carry on reading the file while this variable takes the value `false`. Assuming that we know that the number of points is fewer than 100, we may achieve this using the following code. Note that a potential problem with this code as given will be addressed in Exercise 3.2.

```
 1  #include <cassert>
 2  #include <iostream>
 3  #include <fstream>
 4
 5  int main(int argc, char *argv[])
 6  {
 7    double x[100], y[100];
 8    std::ifstream read_file("Output.dat");
 9    assert(read_file.is_open());
10
11    int i = 0;
12    while (!read_file.eof())
13    {
```

```
14      read_file >> x[i] >> y[i];
15      i++;
16    }
17    read_file.close();
18    return 0;
19 }
```

One additional feature of reading from file that is of use when writing scientific computing applications is the ability to *rewind* a file so that we can read a file starting from the beginning again. This may be achieved by inserting the statements below into the code at the point where the file should be rewound.

```
1    read_file.clear();
2    read_file.seekg(std::ios::beg);
```

3.4 Checking Input and Output are Successful

In Sect. 3.2 we advised C++ programmers to confirm that a file has been opened before writing any data to that file. We justified this using the example of attempting to open a file in a directory that doesn't exist. Under these circumstances the intended data would not be written to file, but the code would proceed without informing us of this.

Even if we do confirm that a file we are intending to write to is open there are other problems that may occur. We may successfully write some data to file, and then reach our disk quota set by the system administrator. Subsequent attempts to write to file would then fail, although the code would continue to execute. Alternatively we may be expecting to read 50 double precision numbers from a file that only contains 40 such numbers. After successfully reading 40 numbers we would like to be informed that we had reached the end of the file, and no more numbers were available to read in. We can check that reading from or writing to file has taken place as expected using the C++ function ios::good. Use of this function is illustrated below for the case of writing to file; its use when reading from file follows a similar pattern.

```
1    std::ofstream write_output("OutputVerified.dat");
2    assert(write_output.is_open());
3    for (int i=0; i<100; i++)
4    {
5      write_output << i << "\n";
6      assert(write_output.good());
7    }
```

3.5 Reading from the Command Line

In scientific computing applications, it is common for a user to want to set some of the parameters used themselves when executing the code. For example, if code has been written to calculate the temperature distribution in a bar using the finite difference method the user may wish to set the thermal conductivity of the bar, or the number of nodes used in the finite difference grid, at the same time that the code is executed. Fortunately, C++ allows the user to do this when running from the command line.

In Sect. 1.2 we promised to explain the third line of the C++ program given in Listing 1.1, namely the line of code shown below.

```
3  int main(int argc, char* argv[])
```

Although we are not quite ready to explain the *whole* meaning of this line until we have introduced pointers in Chap. 4, we may explain how this line allows us to specify input arguments to a program from the command line. Suppose—as described above—we want to write code that allows us to specify an integer number of nodes, number_of_nodes, to be used in a finite difference grid, and a double precision floating point variable, conductivity, that represents the thermal conductivity of a bar. This is demonstrated by the following code. We will explain the additional header file cstdlib used in line 2, and the functions atoi and atof used in lines 15 and 16 at the end of this section: for the time being we will focus on how to input data from the command line.

```
1  #include <iostream>
2  #include <cstdlib>
3
4  int main(int argc, char *argv[])
5  {
6      std::cout << "Number of command line arguments = "
7                << argc << "\n";
8      for (int i=0; i<argc; i++)
9      {
10         std::cout << "Argument " << i << " = " << argv[i];
11         std::cout << "\n";
12     }
13
14     std::string program_name = argv[0];
15     int number_of_nodes = atoi(argv[1]);
16     double conductivity = atof(argv[2]);
17     std::cout << "Program name = " << program_name << "\n";
18     std::cout << "Number of nodes = " << number_of_nodes;
19     std::cout << "\n";
20     std::cout << "Conductivity = " << conductivity << "\n";
21
22     return 0;
23 }
```

We would instruct the user to specify these parameters by typing the executable name, followed by the number of nodes to be used in the finite difference grid, followed by the value for the conductivity: that is, if we want to use 100 nodes and a conductivity of 5.0 we would compile the code above to produce the executable CommandLineCode and then enter the following at the command line:

```
./CommandLineCode 100 5.0
```

This would produce output

```
$./CommandLineCode 100 5.0
Number of command line arguments = 3
Argument 0 = ./CommandLineCode
Argument 1 = 100
Argument 2 = 5.0
Program name = ./CommandLineCode
Number of nodes = 100
Conductivity = 5
$
```

We see from the code and output above that the integer variable argc contains the number of arguments specified at the command line. In this case this is three: these are the executable name ./CommandLineCode, the integer 100, and the floating point number 5.0. These are stored as the ordered list argv[0], argv[1], argv[2], as is demonstrated when we use the for loop to print these out. Each of these are stored as arrays of characters, and so we must first convert these arrays of characters to the appropriate variable types. This is performed by lines 14, 15 and 16 of the code listed. In line 15, we use the function atoi(argv[1]) to convert the array of characters stored by argv[1] to an integer. Similarly, atof(argv[2]) converts argv[2] to a floating point variable. The functions atoi and atof require the header file cstdlib which has been included in line 2.

3.6   Tips: Controlling Output Format

If the files that are written are to be read only by a computer, then it does not really matter whether these look attractive or not. For example, if a data file is only to be used for importing into a visualisation package then it does not matter if the format of this file is opaque to humans provided the visualisation package can read the file accurately. If, however, humans may want to look at these files then formatting commands, such as controlling the width of each column may be desirable.

Below we show how to implement three commonly desired formatting techniques which we now list before demonstrating.

1. *Output in scientific format.* Scientific format is where a number is written as a product of one number with only one significant figure to the left of the decimal point and an integer power of 10, that is, 465.78 in scientific format is 4.6578×10^2, which may be written in C++ notation as 4.6578e2. This is achieved by the use of the flag std::ios::scientific which requires the header file fstream.

2. *Always showing a $+$ or $-$ sign.* The default setting for an output stream is not to print a plus sign before a positive number. To line up numbers in neat columns, we may wish to always precede a number with a plus or minus sign: this is achieved by the use of the flag std::ios::showpos which requires the header file fstream.

3. *Precision of scientific output.* When scientific format is used the precision statement works slightly differently to that described in Sect. 3.2.1: in this case the precision specified is the number of digits *after* the decimal point, and so the number of significant figures is one greater than this number (as there is another significant figure before the decimal point). Furthermore, in contrast to the precision set in Sect. 3.2.1, when scientific format is used zeros are added after the decimal point to ensure that all output is of exactly the same width.

These formatting techniques are demonstrated in the code below.

```cpp
#include <iostream>
#include <fstream>

int main(int argc, char* argv[])
{
    std::ofstream write_file("OutputFormatted.dat");
    // Write numbers as +x.<13digits>e+00 (width 20)
    write_file.setf(std::ios::scientific);
    write_file.setf(std::ios::showpos);
    write_file.precision(13);

    double x = 3.4, y = 0.0000855, z = 984.424;
    write_file << x << " " << y << " " << z << "\n";

    write_file.close();
    return 0;
}
```

3.7 Exercises

3.1 This question assumes that you are starting from the code in the listing below.

```
1  #include <iostream>
2  #include <fstream>
3
4  int main(int argc, char* argv[])
5  {
6      double x[4] = {0.0, 1.0, 1.0, 0.0};
7      double y[4] = {0.0, 0.0, 1.0, 1.0};
8
9      return 0;
10 }
```

1. Extend the code above to print the arrays x and y to a file called x_and_y.dat so that the data file has the four elements of x on the top line, and the four elements of y on the next line.
2. Extend the code so that the output stream is flushed immediately after each line of the file is written.
3. Extend the code so that the precision is set to 10 significant figures, the output is in scientific notation, and plus signs are shown for positive numbers.
4. Amend the program so that it does not automatically create a fresh file x_and_y.dat every time it is run. Have the program first attempt to open the file x_and_y.dat as an ifstream for reading. If the file can be successfully opened then, after closing the ifstream, warn the user. Have the program prompt the user as to whether it should erase the existing file or append to the existing file.

3.2 This question uses the data file x_and_y.dat that was written in the previous exercise. The code below assumes that we know that the data file has 4 columns and that we want to count the number of rows.

```
1  #include <iostream>
2  #include <fstream>
3
4  int main(int argc, char* argv[])
5  {
6      std::ifstream read_file("x_and_y.dat");
7      if (!read_file.is_open())
8      {
9          return 1;
10     }
11     int number_of_rows = 0;
12     while(!read_file.eof())
13     {
14         double dummy1, dummy2, dummy3, dummy4;
15         read_file >> dummy1 >> dummy2;
16         read_file >> dummy3 >> dummy4;
17         number_of_rows++;
18     }
```

```
19    std::cout << "Number of rows = "
20              << number_of_rows << "\n";
21    read_file.close();
22    return 0;
23 }
```

Run the code above. This code does not give the correct answer. Why is this? Does the code give the correct answer if the final newline character is removed from the file x_and_y.dat? Modify the code so that it gives the correct answer. [*Hint: You might investigate the use of* read_file.fail() *which may be used to probe whether the last read on the file stream was unsuccessful.*]

3.3 Write code to implement the implicit (or backward) Euler method to solve the initial value ordinary differential equation

$$\frac{dy}{dx} = -y, \qquad y(0) = 1,$$

on the interval $0 \le x \le 1$ using a constant step size h. Allow the user to specify the number of grid points, N they want to use at the command line, and use an assert statement to ensure that the number of grid points is greater than 1. Use the number of grid points to calculate the step size h. Your code should print a file called xy.dat that has two columns: the calculated values of x; and the calculated values of y. Plot the data from the file xy.dat and hence compare it with the true solution $y = e^{-x}$. [*The implicit Euler method (see, for example, Süli and Mayers* [3]) *for this problem results in the difference relation*

$$y_0 = 1, \qquad \frac{y_n - y_{n-1}}{h} = -y_n, \qquad n = 1, 2, \ldots, N - 1,$$

where h is step size and y_n is the solution at $x_n = nh$, $n = 0, 1, 2, \ldots, N - 1$, where N is the number of grid points, and we have used zero-based indexing for the vectors x *and* y.]

Pointers

4

One of the key features of the C++ language is the concept of a *pointer*. We will see later in this chapter that pointers are extremely useful for allocating memory for arrays whose sizes are not known when the code is compiled. We will see in Chap. 5 that they also have use when writing functions that allow us to repeat the same operation on different variables. We conclude this chapter by discussing some features of pointers that have been introduced in recent C++ standards.

4.1 Pointers and the Computer's Memory

Pointers are best introduced by explaining how they relate to the storage of variables in the computer's memory.

4.1.1 Addresses

Let us suppose that an integer variable total_sum is declared and assigned the value 10:

```
int total_sum = 10;
```

The address—that is, location—of this variable in the computer's memory is given by &total_sum and can be printed to the console in the usual way (as shown below) although this address will not be meaningful to humans.

© Springer International Publishing AG, part of Springer Nature 2017
J. Pitt-Francis and J. Whiteley, *Guide to Scientific Computing in C++*, Undergraduate Topics in Computer Science,
https://doi.org/10.1007/978-3-319-73132-2_4

```
std::cout << &total_sum << "\n";
```

When the integer variable `total_sum` is declared, memory is allocated to this variable, and the location of this memory will not vary throughout execution of the code. As such, the expression `&total_sum`, which represents the address of this location, will take a constant value throughout execution of the code.

4.1.2 Pointer Variables

In addition to data types such as integers and floating point numbers that we have encountered earlier in this book, we may also declare *pointer variables* which are variables that store addresses—that is, the location in the computer's memory—of other variables. In the code below, `p_x` is a pointer to a double precision floating point variable, and `p_i` is a pointer to an integer variable. The pointer `p_x` may then be used to store the address of a double precision floating point number, whilst `p_i` may be used to store the address of an integer. The asterisk that prefixes these variables when they are declared indicates that these variables are pointers. In this book, we follow a coding standard where all pointer variables, apart from those introduced later in this chapter that represent arrays, have names that begin with `p_` to denote that they are a pointer variable: a discussion of conventions such of these that are used for variable names, which forms a part of what is known as *coding standards*, is given in Sect. 6.6.

```
double* p_x;
int* p_i;
```

Note that the spacing can vary, so that `int* p_i` and `int *p_i` are equivalent. However, `int* p_i` states more clearly that the type of `p_i` is a pointer to an integer, rather than an integer.

All pointer variables require an asterisk when they are declared. Hence, in the code below, `p_x, p_y, p_i` are pointers, while `j` is an integer variable.

```
double *p_x, *p_y;
int *p_i, j;
```

When declaring more than one pointer on a line the asterisk must be repeated as shown in line 1 of the listing above, which means that `int* p_i` in line 2 would be less appropriate as only one variable (`p_i`) is a pointer variable. For this reason, we recommend only one pointer declaration per line.

Now we have explained how to declare a pointer variable, and what these variables represent, we explain how to use them.

4.1.3 Example Use of Pointers

If a variable p_x has been declared as a pointer to a double precision floating point number, then it is clearly important to distinguish between: (i) the location of the memory to which this pointer points at (denoted by p_x); and (ii) the contents of this memory (denoted by *p_x). The asterisk operator in *p_x is called a *pointer de-reference* and can be thought of as the opposite to the & operator introduced in Sect. 4.1.1.

The code below shows how pointers to double precision floating point variables may be combined with double precision floating point variables.

```
1   double y, z;       // y, z store double precision numbers
2   double* p_x;       // p_x stores the address of a double
3                      // precision floating point number
4   z = 3.0;
5   p_x = &z;          // p_x stores the address of z
6   y = *p_x + 1.0;    // *p_x is the contents of the memory
7                      // p_x, i.e. the value of z
```

4.1.4 Warnings on the Use of Pointers

A variable pointer should not be used until first having been assigned a valid address. For example, the following fragment of code may cause problems that are difficult to locate.

```
1   double* p_x; // p_x can store the address of a double
2                // precision number - haven't said which
3                // address yet
4
5   *p_x = 1.0; // trying to store the value 1.0 in an
6                // unspecified memory location
```

In the code above, we haven't specified the location of the double precision floating point variable that p_x points at. It may therefore be pointing at *any* location in the computer's memory. Changing the contents of an unspecified location in a computer's memory—as is done in line 5 of the code above—clearly has the potential to cause problems that may be hard to locate. This problem may be avoided by the use of the new keyword as shown below to allocate a valid memory address to p_x, and the delete keyword which releases this memory to be used by other parts of the program when this memory is no longer required.

```
1   double* p_x;        // p_x stores the address of a double
2                       // precision floating point number
3
4   p_x = new double;   // assigns an address to p_x
5   *p_x = 1.0;         // stores 1.0 in memory with
6                       // address p_x
7   delete p_x;         // releases memory for re-use
```

A further reason to use pointers with care is shown in the code below. The first time y is printed (in line 5) it takes the value 3: the second time y is printed (in line 7) it takes the value 1 even though y is not explicitly altered in the code between these two lines. This is because the line between the std::cout statements, line 6, has altered the value of y, possibly unintentionally, by using the pointer variable p_x (which contains the address of y) to change the value of y.

```
1   double y;
2   double* p_x;
3   y = 3.0;
4   p_x = &y;
5   std::cout << "y = " << y << "\n";
6   *p_x = 1.0;  // This changes the value of y
7   std::cout << "y = " << y << "\n";
```

A situation where the contents of the same variable may be accessed using different names, such as in the code above, is known as *aliasing*. In C++, this is most likely to happen when pointers are involved, either when two pointers alias the same address in memory, or when a pointer references the contents of another variable. When one or more pointers allow the same variable to be accessed using different names, the aliasing is known as *pointer aliasing*.

4.2 Dynamic Allocation of Memory for Arrays

One of the main uses of pointers is the dynamic allocation of memory for storing arrays. In Sect. 1.4.5, we explained how arrays could be declared when the size of the array was known in advance. However, we do not always know the sizes of the arrays in a program when we compile the code. In Sect. 3.5, for example, we demonstrated how to allow the user of a code to specify the number of nodes in a finite difference grid when executing the code. If the coordinates of the nodes in this mesh were to be stored in an array we would not know, when compiling the code, what size to make this array. Under these circumstances, using the method of declaring arrays given in

Sect. 1.4.5, we have to compile the code with some estimate of the size of this array. If we overestimate the size of this array, we are being wasteful of computational memory with the potential effect of preventing the execution of the code on a system with insufficient memory. If we underestimate the size of this array, the program will almost certainly crash. In either case, we will then have to recompile the code with a new estimate of the array size. The use of pointers to dynamically allocate memory for arrays avoids these problems, as we do not need to know the array size at compile time.

A further use of pointers for dynamically allocating memory is for the efficient storage of irregularly sized arrays, for example a lower triangular matrix. If a lower triangular matrix is stored in an array as described in Sect. 1.4.5, we will have to allocate the same number of columns to each row of the matrix. As we know that roughly half these entries are zero, we are being wasteful of computational memory. Dynamic allocation of memory allows us to allocate memory more prudently.

Memory can be allocated using the `new` operator, and deallocated using the `delete` operator.

4.2.1 Vectors

To use pointers to create a one-dimensional array of double precision floating point numbers of length `10` called `x`, we use the following section of code.

```
double* x;
x = new double [10];
```

The elements of the array may then be accessed in exactly the same way as if the array had been created by using the type of declaration introduced in Sect. 1.4.5. In the dynamic allocation of memory for the array using the pointer `x` above, `x` stores the address of the first element of the array. This can be seen by printing out both the pointer `x` and the address of the first element of the array, as shown below.

```
std::cout << x << "\n";
std::cout << &x[0] << "\n"; //prints the same value
```

The memory allocated to `x` may be, and should be, deallocated by using the statement below when this array is no longer required.

```
delete[] x;
```

Always be sure to free any memory allocated when it is no longer required—a code can very quickly use all available memory otherwise. In later chapters of this book, when we develop a class of vectors, we will see that one advantage of writing a class of vectors is that memory allocated to a vector is automatically freed when appropriate.

An example code that uses dynamically allocated memory for arrays is shown below. This code creates two arrays, x and y, both of size 10. Elements of x are then assigned manually. Elements of y are then set to be twice the value of the corresponding element of x. Finally, all memory allocated is deleted.

```cpp
#include <iostream>

int main(int argc, char* argv[])
{
    double* x;
    double* y;
    x = new double [10];
    y = new double [10];

    for (int i=0; i<10; i++)
    {
        x[i] = ((double)(i));
        y[i] = 2.0*x[i];
    }

    delete[] x;
    delete[] y;

    return 0;
}
```

4.2.2 Matrices

Memory for matrices may also be allocated dynamically. For example, to create a two-dimensional array of double precision floating point numbers with 5 rows and 3 columns called A we use the following section of code.

Listing 4.1 Dynamic memory allocation for a matrix

```cpp
int rows = 5, cols = 3;
double** A;
A = new double* [rows];
for (int i=0; i<rows; i++)
{
    A[i] = new double [cols];
}
```

The array may then be used in exactly the same way as if it had been created by using the declaration

```
double A[5][3];
```

When allocating memory for the matrix dynamically in the code above, the variable A—which has been declared using line 2 of Listing 4.1—has the following properties after the fragment of code has been executed:

- each A[i] is a pointer, and contains the address of A[i][0]; and
- A contains the address of the pointer A[0].

As such, the variable A is an array of pointers, which explains the two asterisks in line 2 of Listing 4.1. Line 3 of this listing specifies that A is a pointer to an array of pointers to double precision floating point numbers, and that this array is of size rows. The for loop in this listing then specifies that each pointer in the array itself points to an array of double precision floating point numbers of length cols. This has the effect that A[i]—which is a pointer—stores the address of the entry A[i][0], that is, the first entry of row i.

As was the case for vectors, it is important to deallocate memory dynamically allocated for a matrix when it is no longer needed. The memory allocated for the matrix A in Listing 4.1 may be freed using the following code.

```
for (int i=0; i<rows; i++)
{
    delete[] A[i];
}
delete[] A;
```

We cannot emphasise enough how important it is to always delete any memory dynamically allocated, particularly memory allocated inside loops—if not you will soon run out of memory.

4.2.3 Irregularly Sized Matrices

Suppose we want to construct a lower triangular matrix A of integers with 1,000 rows and 1,000 columns. This may clearly be done using the declaration below.

```
int A[1000][1000];
```

However, the declaration above wastes a considerable amount of memory storing the super-diagonal entries of the matrix which we know in advance all take the value 0. We may avoid wasting this memory by allocating the memory for this matrix dynamically, and only allocating memory for the diagonal and sub-diagonal elements. This is demonstrated in the fragment of code below, where in row i of the matrix we declare i+1 nonzero elements: that is, 1 element in row 0, 2 elements in row 1, and so on. Memory can, and should be, deleted in the same way as demonstrated in the previous section when this array is no longer required.

```
1    int** A;
2    A = new int* [1000];
3    for (int i=0; i<1000; i++)
4    {
5        A[i] = new int[i+1];
6    }
```

Although the fragment of code above does correctly allocate the memory required for a lower triangular matrix it should be used with care: errors would result if, for example, the entry A[9][19] were to be used in a code. When we develop classes later in this book, we will see how the use of classes may avoid problems such as this.

4.3 Tips: Pointers

The concept of pointers is one that inexperienced C++ programmers often struggle with. We strongly urge the reader to attempt the exercises at the end of this chapter to improve their understanding of this topic. In this section, we give tips on the use of pointers. In all other chapters the tips section is the final section before the exercises. This chapter is the exception because some of the caution in the following tips may be mitigated in modern C++. We introduce these advanced topics in Sect. 4.4 which you might ignore on first reading.

4.3.1 Tip 1: Pointer Aliasing

In Sect. 4.1.4, we gave an example where a pointer variable p_x was pointing to the memory location of the double variable y. A change was made to that variable by de-referencing the pointer p_x. This situation might lead to some confusion, although in a short code fragment it is easy to see that the two variables are leading to the same place: *p_x is an *alias* for y.

In large-scale programs, it may not be so easy to see where pointers are aliases for other variables. This is because the information that two names are pointing to same place may not be available in the same screen-full of code, or even in the same file. A good example of this would be a vector or matrix addition operation in which the vectors or matrices are stored as arrays and passed into a *function* via pointers. We will deal with functions in the next chapter, but for now you need to be aware that the code for the function may be in a different file and that the variables may take different names inside the function definition. The operation to compute the matrix sum $\mathbf{A} = \mathbf{B} + \mathbf{C}$ would probably be implemented in such a function by a nested loop over the elements of the arrays, so that the actual implementation becomes an element-wise `A[i][j] = B[i][j] + C[i][j]`. There may be unknown pointer-aliasing in this function, because the user might wish to increment one matrix by another, i.e. to compute $\mathbf{X} = \mathbf{X} + \mathbf{Y}$. It turns out that this pointer aliasing will be safe, because the inner loop will effectively be calculating `X[i][j] += Y[i][j]` as intended. Each of the (i, j) components of the result is independent of the others.

However, what if the user were using a matrix–matrix product operation? In the computation $\mathbf{A} = \mathbf{BC}$, the component `A[i][j]` depends on parts of $\mathbf{B}$ and $\mathbf{C}$ other than `B[i][j]` and `C[i][j]`. This means that, if the user wishes to compute $\mathbf{X} = \mathbf{XY}$ using a function written for calculating $\mathbf{A} = \mathbf{BC}$, there is a chance that some components of $\mathbf{X}$ will be written to before they are read—leading to an incorrect calculation. One way to resolve this aliasing issue is to produce the matrix–matrix product result in temporary storage before copying it into the output argument $\mathbf{A}$. However, this solution is inefficient in cases where there is no pointer aliasing, especially when the sizes of the matrices are large. Another solution to the issue is to provide two versions of the matrix-matrix product operation: one which is efficient but only safe to use when there is no pointer aliasing and one which is safe to use in all circumstances.

One can see that the problem of pointer aliasing is deeper than might appear from the trivial example in Sect. 4.1.4. In general, there is no correct solution to these issues. Compiler writers spend a great deal of time finding places where pointer aliasing has (or has not) definitely happened so that code optimisation is only applied in situations where it is safe to do so.

4.3.2 Tip 2: Safe Dynamic Allocation

There may be circumstances under which it is not possible to allocate memory either because the number of items in an array has been set with a negative argument or because there is not enough physical memory available to the program. Setting the number of elements in an array to a negative number is easier than you might think. If the size of a problem is configured via an input file, then a size may easily be mistyped. More subtly, if a number is assigned to an integer that is larger than the maximum value that can be stored by that integer, then the integer value stored may actually be a negative number: this is known as an *overflow error*.

Implementations of C++ may vary over how they treat such errors. The default behaviour is to throw an *exception* when a memory error is encountered. We will deal with catching exceptions in Chap. 9 and note that an exception could terminate your program. Should your implementation of C++ not throw this sort of exception, then a safe way to program is to test that your variable has been assigned a value as the code fragment below illustrates.

```
1    double* p_x;
2    p_x = new double[10000];
3    assert (p_x != NULL);
```

4.3.3 Tip 3: Every new Has a delete

We pointed out earlier in this chapter that all dynamically allocated memory must be freed, or else you may run out of memory. This problem is particularly noticeable when memory is dynamically allocated inside the body of a `for` loop, such as the one shown below.

```
1    for (int i=0; i<10000; i++)
2    {
3        double** A;
4        A = new double* [50];
5        for (int j=0; j<50; j++)
6        {
7            A[j] = new double [50];
8        }
9    }
```

Each time the body of the loop in the code above is executed, new memory is allocated for the array A. The memory from the previous execution has not yet been freed, although it will not be available as the array A will be stored in the memory that has been allocated most recently: there is no automatic garbage collection for memory which is no longer accessible. You will see, when we discuss functions in Chap. 5, that the same problem may arise when memory is allocated inside functions, but not freed before the function ends.

If you do not delete memory which you have allocated dynamically, then that memory will not be accessible until your program finishes (when all memory is handed back to the system). If you request more memory than you need, then it may be that the physical memory of the computer will be exhausted—your computer will run much more slowly and further memory allocation may fail.

There are several ways around this issue. The first and foremost is to ensure that every new in your program is matched with a delete somewhere else. A second way to make sure that inaccessible or unnecessary memory is freed up is to run your program through a memory debugger (see Sect. 10.6 for more details). Another solution, adopted by seasoned C++ programmers is to use *shared pointers*. These are an advanced language feature which allow memory to be automatically de-allocated once there is no longer any other part of the program which can access it.

4.4 Modern C++ Memory Management

In Sect. 1.1.2, when discussing why you should write scientific programs in C++, we claimed that its flexible memory management gave it an advantage over languages which use garbage collection, such as Java. However we also gave a caveat: this flexible memory management means that you, the programmer, are responsible for making sure that memory is managed properly. Many novice C++ programmers are confused by dynamic memory allocation and become deterred when they learn that it is up to them to know when dynamically created data should be freed up with `delete`. The good news for C++ programmers is that over recent years the C++ standard has introduced smart pointer constructs which facilitate memory management—providing an efficient compromise between giving responsibility to the programmer and automatic run time garbage collection. These constructs were first introduced in the C++11 specification and have been refined in subsequent specifications.[1] In this chapter we restrict attention to modern C++ *memory management* but we will return to other modern C++ functionality in Chap. 8.

4.4.1 The `unique_ptr` Smart Pointer

In our first tip of this chapter, in Sect. 4.3.1, we warned about the dangers of pointer aliasing. In particular we noted that there may be times when a programmer assumes that two pointers are pointing to different pieces of data, but that this assumption may not be true. When two pointers are pointing to the same piece of data it may lead to bugs such as an element of a matrix being overwritten before its value has been read.

C++11 provides a smart pointer type which can guard against pointer aliasing errors. This smart pointer `unique_ptr` allows the run-time system to monitor certain pointers on an individual basis. The example of its use, given in Listing 4.2, is a little contrived because the true power of the construct cannot be seen until it is used with functions. The program will, however, serve to illustrate a few of the main features. Your C++ compiler may not accept this program since most current compilers are set to read older C++98 standard programs by default. In order to compile the program you will need to add a flag to indicate that the code adheres to the C++11 standard. In the case of the GNU compiler this means

```
g++ -std=c++11 -o Unique Unique.cpp
```

or similar.

In line 6 of Listing 4.2 a new `int` is dynamically created via the `new` keyword and its address assigned to a `unique_ptr` called `p_x`. Note that the type description of

---

[1] At the time of writing the second edition of this book the relevant specifications are C++11, C++14 and C++17.

the unique_ptr contains the type of the entity to which it points, which in this case is int, in angle brackets. This angle bracket notation is a *template* description and we will see more of this in Chap. 8. The purpose of new int in round brackets on line 6 is to dynamically create an int and pass its location into p_x. The variable p_x now acts as a facade through which the actual address of the dynamically-created integer storage may be accessed. There is more happening behind the scenes, but the reader may interpret line 7 as a de-reference used to store a value in the memory location at this address.

We demonstrate that the compiler won't allow us to easily assign the value of p_x by two lines which have been commented out: line 11 attempts to assign it to a raw pointer and line 16 attempts to assign it to another unique_ptr. The correct way to get the value out of p_x (line 12) is to use the get() function to get the actual address of the managed data. Meanwhile the correct way to assign from one unique_ptr to another is for the ownership of the resource to be transferred between them with the function std::move(). This is demonstrated in line 17. Lines 18 and 19 show that the unique_ptr variables can be evaluated as Boolean values: true if the variable is managing a resource and false if not.

Note that in Listing 4.2 there is no explicit call to delete to match with the new on line 6. It is actually the case that, because the unique_ptr is managing the resource, it is able to automatically free up memory. On line 20, p_z is told to relinquish ownership and this implicitly calls delete on the memory originally created on line 6.

Listing 4.2 Example program to demonstrate the use of unique_ptr

```
1   #include <memory> // Requires C++11 or above
2   #include <cassert>
3
4   int main()
5   {
6     std::unique_ptr<int> p_x(new int);
7     *p_x = 5; // 'de-reference' to alter contents
8
9     // The following won't compile because p_x
10    // is not a raw pointer to int
11    //   int* p_y = p_x;
12    int* p_y = p_x.get(); // Get raw pointer
13
14    std::unique_ptr<int> p_z;
15    // The following won't compile
16    // p_z = p_x;
17    p_z = std::move(p_x); // Transfer ownership
18    assert(p_z); // Test p_z is in use
19    assert(!p_x); // Test that p_x not in use
20    p_z.reset();
21    assert(!p_z); // Test p_z is also not in use
22    return 0;
23  }
```

4.4.2   The `shared_ptr` Smart Pointer

The mismatch between the last code example (Listing 4.2) and our previously sound advice in Sect. 4.3.3, "Every `new` has a `delete`", prompts us to introduce the variable type `shared_ptr`. This smart pointer construct was not available in the official C++ standard until C++14 but some C++11 compilers such as the GNU compiler support it anyway.

The concept behind a smart shared pointer is simple. Alongside the address of the underlying resource the pointer also keeps track of a count of the number of times this resource has been used. Initially the count will be 1, but it will increment when the pointer is passed between various parts of the program. Whenever a use of the pointer finishes the usage count will be decremented. When the count drops to 0, and there are no known uses of the pointer, the original resource will be freed up. This all happens automatically, without the user having to worry about it. It is effectively a local garbage collector which manages a small piece of memory.

The code presented in Listing 4.3, illustrating the use of a smart shared pointer, is again a little contrived, but it represents how this automatic memory management might work in practice. In line 6 a new integer value is dynamically created and its location is stored in a `shared_ptr` variable `p_x`. As with the previous C++11 example this new smart pointer is templated with the type of its argument in angle brackets. In line 10 another `shared_ptr` variable is created and it is assigned to the same value as `p_x` (this is an assignment which would not be possible with the `unique_ptr` type). In line 12 `p_y` is reset so that it relinquishes any claim on memory. While lines 10 and 12 are just a simple assignment and a reset, respectively, their use here actually represents general wider uses of a shared pointer. Copies of pointers may be made when they are passed into functions, as we will see in Chap. 5, or passed into *containers*—of the kind introduced in Chap. 8. When functions or containers finish, their copy of the pointer is not needed and is, in effect, reset.

Listing 4.3 Example program to demonstrate the use of `shared_ptr`

```
1   #include <memory> // Requires C++11 or above
2   #include <iostream>
3
4   int main()
5   {
6       std::shared_ptr<int> p_x(new int);
7       std::cout<<"p_x use count: "<<p_x.use_count()<<"\n";
8       *p_x = 5; // 'de-reference' to alter contents
9       // Use this pointer elsewhere
10      std::shared_ptr<int> p_y = p_x;
11      std::cout<<"p_x use count: "<<p_x.use_count()<<"\n";
12      p_y.reset();
13      std::cout<<"p_x use count: "<<p_x.use_count()<<"\n";
14      p_x.reset();
15      std::cout<<"p_x use count: "<<p_x.use_count()<<"\n";
16      return 0;
17  }
```

Lastly in line 14 the original pointer is reset. This has, again, the same effect as p_x going out of use: its claim on the data is relinquished. In this case the use count will drop to 0 and the smart pointer will automatically free up the original memory which was created on line 6.

Throughout Listing 4.3 the use count of the main shared pointer p_x is written to the console. The output of this program is given below and reflects the number of uses of the shared resource. This count is originally 1 when p_x is created, then 2 when p_y shares the resource, and 1 when p_y relinquishes its use on line 12. Finally, when p_x relinquishes its use, the count drops to 0.

```
p_x  use  count: 1
p_x  use  count: 2
p_x  use  count: 1
p_x  use  count: 0
```

4.5 Exercises

4.1 Write code that declares an integer i to take the value 5. Declare a pointer to an integer p_j, and store the address of i in this pointer. Multiply the value of the variable i by 5 by using a line of code that *only uses the pointer variable*. Declare another pointer to an integer p_k and use the new keyword to allocate a location in memory that this pointer stores. Then store the contents of the variable i in this location. Now change the value pointed to by p_j to 0. Check that your program is correct by outputting the value of i and values pointed to by p_j and p_k.

4.2 Assign values to two integer variables. Swap the values stored by these variables using only pointers to integers.

4.3 Write code that allocates memory dynamically to two vectors of double precision floating point numbers of length 3, assigns values to each of the entries, and then de-allocates the memory before the code terminates. Extend this code so that it calculates the scalar (dot) product of these vectors and prints it to screen before the memory is de-allocated. Put the allocation of memory, calculation and de-allocation of memory inside a for loop that runs 1,000,000,000 times: if the memory is not de-allocated properly your code will use all available resources and your computer may struggle.

4.4 Write code that dynamically allocates memory for three 2×2 matrices of double precision floating point numbers, A, B, C, and assigns values to the entries of A and B. Let C = A + B. Extend your code so that it calculates the entries of C, and then prints the entries of C to screen. Finally, de-allocate memory. Again, check you have de-allocated memory correctly by using a for loop as in the previous exercise.

4.5 In Sect. 4.4 we introduced the `unique_ptr` and `shared_ptr` constructs. A useful further smart pointer is the `weak_ptr`, which is a smart pointer that does not contribute to the use count. It can be used in situations where variables need to be accessed, but only when they exist. It has functions `expired` and `lock` which can be used to check if its resource has been deleted and, if it has not been deleted, to get to the resource.

Copy Listing 4.3 and compile it with a compatible C++11 compiler. Now add an extra smart pointer: a `weak_ptr` which is initialised to the value `p_x`. Experiment with printing the value original of `p_x` (i.e. the value 5) via this weak smart pointer. Try this before, and after, the `p_x` is reset on line 14.

Blocks, Functions and Reference Variables

The code developed in this book up to this point has been restricted to code that may be placed inside curly brackets after the initial line of code "int main(int argc, char* argv[]);". Readers with previous programming experience will be aware of the limitations this places when writing code. For example, if we were to apply the same operations in different places in the code we would have to repeat the lines of code that performed these operations everywhere in the code where they were required. It would be much more convenient if we could write a function that we could call whenever we wanted to perform these operations. This chapter introduces the C++ machinery for writing functions.

5.1 Blocks

A block is any piece of code between curly brackets. A variable, when declared inside a block, may be used throughout that block, but only within that block. This is demonstrated in the code below. In line 9, we attempt to use the variable j when it is only declared—and therefore available—in the block enclosed within the curly brackets in lines 4 and 8. In the language of programmers, "the scope of j is the block between lines 4 and 8". If we attempted to use the code fragment below, the compiler would report this attempted use of j as an error: j is said to be *out of scope* at line 9.

© Springer International Publishing AG, part of Springer Nature 2017

J. Pitt-Francis and J. Whiteley, *Guide to Scientific Computing in C++*, Undergraduate Topics in Computer Science, https://doi.org/10.1007/978-3-319-73132-2_5

```
1      {
2          int i;
3          i = 5;   // OK
4          {
5              int j;
6              i = 10; // OK
7              j = 10; // OK
8          }
9          j = 5;   // incorrect - j not declared here
10     }
```

The same variable name may be used for a variable declared both inside a block—termed the *local variable*—and outside the scope of any function (including the main function)—termed the *global variable*. Both of these variables may be accessed inside the block as shown in the code below, using the example of both a global and a local variable called i. Furthermore, we may define a variable j in both the outer block and the inner block: inside the inner block the value of j stored by the variable declared in the outer block is not accessible. The multiple declaration of both i and j in the code below is bad programming practice, as it can clearly lead to confusion. In fact, since the scope of variables is so important, we suggest that variables are declared only within the block where they are needed, close to their first use. This multiple declaration of variables is known as *variable shadowing* and you can avoid it happening by turning on "shadow warnings" in your compiler. With the GNU g++ compiler this is achieved by adding the -Wshadow flag to the compilation command.

```
1   #include <iostream>
2   int i = 5; // global i
3
4   int main(int argc, char* argv[])
5   {
6       int j = 7;
7       std::cout << i << "\n";
8       {
9           int i = 10, j = 11;
10          std::cout << i << "\n"; // local value of i is 10
11          std::cout << ::i << "\n"; // global value of i is 5
12          std::cout << j << "\n"; // value of j here is 11
13          //The other j (value 7) is inaccessible
14      }
15      std::cout << j << "\n"; // value of j here is 7
16      return 0;
17  }
```

5.2 Functions

Now that we have defined what we mean by a block of code we may demonstrate how to write functions.

5.2.1 Simple Functions

A simple program that writes and uses a function to determine the minimum value of two double precision floating point variables x and y, and stores it in the double precision variable minimum_value is shown below. Note the *function prototype* that is line 3 in the listing below. The function prototype tells the compiler what input variables are required, and what variable, if any, is returned. In the example below, the function prototype explains that later in the code there will be a function called CalculateMinimum that requires two double precision floating point variables as input, and returns one double precision floating point variable. The function prototype can be thought of as being similar to declaring a variable. The variable names a and b in the prototype are ignored by the compiler and don't have to be included, but their inclusion can clarify the program. Note that the function prototype ends with a semi-colon.

   Lines 15–29 of the code contain the statements that perform the tasks required by the function. This code begins with a line of code that is identical to the function prototype (including the variable names) without the semi-colon. After this there is a block of code that ends with a return statement that returns the value required to the point in the code where this function was called from. Note that there is no need to declare the variables a and b inside the function—the declaration in line 15 has done this already. Variables such as minimum that are used inside the function but are not part of the function prototype must be declared within the function block. Line 8 demonstrates how to call a function: the variables in brackets (x and y in this case) are sent to the function, and are known as the *arguments* of the function. The variable returned from the function is stored as minimum_value.

```
1   #include <iostream>
2
3   double CalculateMinimum(double a, double b);
4
5   int main(int argc, char* argv[])
6   {
7       double x = 4.0, y = -8.0;
8       double minimum_value = CalculateMinimum(x, y);
9       std::cout << "The minimum of " << x << " and " << y
10                 << " is " << minimum_value << "\n";
11
12      return 0;
13  }
14
```

```
15   double CalculateMinimum(double a, double b)
16   {
17      double minimum;
18      if (a < b)
19      {
20         minimum = a;
21      }
22      else
23      {
24         // a >= b
25         minimum = b;
26      }
27
28      return minimum;
29   }
```

Note that only one variable may be returned from a function. Although sufficient for some purposes, we may sometimes want to return more variables. We will see how this may be done later in this chapter. Of course, there are some circumstances where we do not want a function to return any variable: such functions may be prototyped as a void function. The code below contains an example of a function that prints out a message informing a candidate whether or not they have passed an exam. This function requires two integer variables as input: the first of these contains the mark that a candidate has scored; the second contains the pass mark for the exam.

```
1    #include <iostream>
2
3    void PrintPassOrFail(int score, int passMark);
4
5    int main(int argc, char* argv[])
6    {
7       int score = 29, pass_mark = 30;
8       PrintPassOrFail(score, pass_mark);
9
10      return 0;
11   }
12
13   void PrintPassOrFail(int score, int passMark)
14   {
15      if (score >= passMark)
16      {
17         std::cout << "Pass - congratulations!\n";
18      }
19      else
20      {
21         // score < passMark
22         std::cout << "Fail - better luck next time\n";
23      }
24   }
```

A function can only change the value of a variable sent to a function *inside* that function: changes made within the function will have no effect on this variable after the function has been executed and the code continues to execute statements in the block where the function has been called from. This is because a copy is made of any variable that is sent to a function, and it is this copy of the variable, and not the original variable, that is modified inside the function. For example, the following function has no effect on the variable x outside the function, even though the value of x *is* changed inside the function.

```
#include <iostream>

void HasNoEffect(double x);

int main(int argc, char* argv[])
{
    double x = 2.0;
    HasNoEffect(x);
    std::cout << x << "\n";   // will print out 2.0

    return 0;
}

void HasNoEffect(double x)
{
    // x takes the value 2.0 here
    x += 1.0;
    // x takes the value 3.0 here
}
```

5.2.2 Returning Pointer Variables from a Function

In Sect. 5.2.1, we demonstrated how to write functions that returned either a variable that wasn't a pointer, or had no return type. Functions can be used to return pointer variables as well, as shown in the code below. In this case, we have written a function that allocates memory for a matrix dynamically, and returns the pointer to the memory allocated. The array can then be used as if the memory were allocated in the main function, as demonstrated in lines 8 and 9. Because every new requires a matching delete we have avoided leaking memory by also providing a function called FreeMatrixMemory to free up the memory created in AllocateMatrixMemory. Both AllocateMatrixMemory and its partner function FreeMatrixMemory operate in the manner described in Sect. 4.2.2.

```
1   double** AllocateMatrixMemory(int numRows, int numCols);
2   void FreeMatrixMemory(int numRows, double** matrix);
3
4   int main(int argc, char* argv[])
5   {
6     double** A;
7     A = AllocateMatrixMemory(5, 3);
8     A[0][1] = 2.0;
9     A[4][2] = 4.0;
10    FreeMatrixMemory(5, A);
11    return 0;
12  }
13
14  // Function to allocate memory for a matrix dynamically
15  double** AllocateMatrixMemory(int numRows, int numCols)
16  {
17    double** matrix;
18    matrix = new double* [numRows];
19    for (int i=0; i<numRows; i++)
20    {
21      matrix[i] = new double [numCols];
22    }
23    return matrix;
24  }
25
26  // Function to free memory of a matrix
27  void FreeMatrixMemory(int numRows, double** matrix)
28  {
29    for (int i=0; i<numRows; i++)
30    {
31      delete[] matrix[i];
32    }
33    delete[] matrix;
34  }
```

5.2.3 Use of Pointers as Function Arguments

We concluded Sect. 5.2.1 by explaining that any changes to a variable made inside a function would have no effect outside that function. This has the advantage that if a variable is altered unintentionally then the impact of this is localised to the function where this unintentional alteration was made. However, there are occasions where we *do* wish changes to a variable inside a function to have an effect outside a function. For example, if we are given a complex number in polar form, $z = re^{i\theta}$, we may wish to write a function that returns the real part, denoted by the variable x, and imaginary part, denoted by the variable y, of this number. We have noted earlier that a function can only return one variable, and so we may not return both the variable x and the variable y. It would therefore be useful to include the variables x and y in the function call. However, this would not work either, as the values assigned to these variables would not have any effect outside the function. Fortunately pointers provide us with one way around this problem. Instead of sending the variables x and y to the function, we send the *addresses* of these variables to the function. When the function is called, copies are made of the addresses of these variables, and it is these copies that are sent to the function. Changes to these addresses will not have any effect outside the function as we are working with a copy of these addresses. However, we can change the contents of the variable without changing the address through de-referencing the pointer, and this will have an effect outside of the function. This is demonstrated in the code below.

Note that lines 4–6 of the code are really meant to be one long line, giving the function prototype of CalculateRealAndImaginary. Since the line is long, we have split it across several lines and indented the continuation lines for clarity (see Sect. 6.6 for a discussion of stylistic conventions when writing code). The prototype lists the arguments for the function. The first two arguments are double precision floating point variables representing the magnitude (denoted by r) and argument (denoted by theta) of the specified complex number. The third and fourth arguments are pointers to—that is, the addresses of—the real part and imaginary part of the complex number. In line 12, we declare integers x and y that represent the real and imaginary parts of the complex number. To use the function CalculateRealAndImaginary, we send the addresses of these variables to the function. Behind the scenes a copy of these addresses is made, and it is these copies that are used in the function in lines 20–26. However, these copies refer to the same memory as the original variables x and y, and so it is this memory that the results of the calculations in lines 24 and 25 are stored in.

Listing 5.1 Use of pointers with functions

```
1   #include <iostream>
2   #include <cmath>
3
4   void CalculateRealAndImaginary(double r, double theta,
5                                  double* pReal,
6                                  double* pImaginary);
7
8   int main(int argc, char* argv[])
9   {
10      double r = 3.4;
11      double theta = 1.23;
12      double x, y;
13      CalculateRealAndImaginary(r, theta, &x, &y);
14      std::cout << "Real part = " << x << "\n";
15      std::cout << "Imaginary part = " << y << "\n";
16
17      return 0;
18  }
19
```

```
20   void CalculateRealAndImaginary(double r, double theta,
21                                    double* pReal,
22                                    double* pImaginary)
23   {
24      *pReal = r*cos(theta);
25      *pImaginary = r*sin(theta);
26   }
```

5.2.4 Sending Arrays to Functions

When sending arrays to functions—whether or not the memory has been allocated dynamically—it should be noted that it is the address of the first element of the array that is being sent to the function. In common with sending the pointer to a variable to a function, changes to this address will not have an effect in the code from which this function is called: however, the contents of this address—that is, the contents of the array—may be changed. As such, any changes made to an array inside a function *will* have an effect when that variable is used subsequently outside the function.

We begin by showing how to send arrays whose size is known at compile time to a function. This is shown in the listing below. Note that we do not have to specify the size of the first index of an array in the function prototype. This size is computed by the compiler. It may be included if desired, but this will be ignored when the code is compiled.

```
1    #include <iostream>
2    #include <cmath>
3
4    void DoSomething(double u[], double A[][10],
5                     double B[10][10]);
6
7    int main(int argc, char* argv[])
8    {
9       double u[5], A[10][10], B[10][10];
10
11      DoSomething(u, A, B);
12
13      // This will print the values allocated in
14      // the function DoSomething
15      std::cout << u[2] << "\n";
16      std::cout << A[2][3] << "\n";
17      std::cout << B[3][3] << "\n";
18
19      return 0;
20   }
21
22   void DoSomething(double u[], double A[][10],
23                    double B[10][10])
```

```
24  {
25      u[2] = 1.0;
26      A[2][3] = 4.0;
27      B[3][3] = -90.6;
28  }
```

Arrays whose size has been dynamically allocated can also be sent to a function. Example code for this is shown below.

```
1  #include <iostream>
2  #include <cmath>
3
4  void DoSomething(double* u, double** A);
5
6  int main(int argc, char* argv[])
7  {
8      double* u = new double [10];
9      double** A = new double* [10];
10     for (int i=0; i<10; i++)
11     {
12         A[i] = new double [10];
13     }
14
15     DoSomething(u, A);
16
17     // This will print the values allocated in
18     // the function DoSomething
19     std::cout << u[2] << "\n";
20     std::cout << A[2][3] << "\n";
21
22     delete[] u;
23     for (int i=0; i<10; i++)
24     {
25         delete[] A[i];
26     }
27     delete[] A;
28
29     return 0;
30 }
31
32 void DoSomething(double* u, double** A)
33 {
34     u[2] = 1.0;
35     A[2][3] = 4.0;
36 }
```

5.2.5 Example: A Function to Calculate the Scalar Product of Two Vectors

Suppose we want to calculate the scalar product of two vectors of double precision floating point numbers of length n. Calculating the scalar product could be embedded within a function that inputs the two arrays, and the length n of both vectors, and returns a double precision floating point variable that represents the scalar product of the two vectors: see Sect. A.1.2 for a discussion of how to calculate the scalar product of two vectors. We would first need to allocate memory for the two vectors. We could then call the function that calculates the scalar product, before finally deleting the memory allocated to the two vectors. Code for this is shown below.

```
1   #include <iostream>
2
3   double CalculateScalarProduct(int size, double* a,
4                                 double* b);
5
6   int main(int argc, char* argv[])
7   {
8      int n = 3;
9      double* x = new double [n];
10     double* y = new double [n];
11     x[0] = 1.0;   x[1] = 4.0;   x[2] = -7.0;
12     y[0] = 4.4;   y[1] = 4.3;   y[2] = 76.7;
13     double scalar_product = CalculateScalarProduct(n, x, y);
14     std::cout << "Scalar product = "
15               << scalar_product << "\n";
16     delete[] x;
17     delete[] y;
18
19     return 0;
20  }
21
22  double CalculateScalarProduct(int size, double* a,
23                                double* b)
24  {
25     double scalar_product = 0.0;
26     for (int i=0; i<size; i++)
27     {
28        scalar_product += a[i]*b[i];
29     }
30     return scalar_product;
31  }
```

5.3 Reference Variables

In Sect. 5.2.3, we demonstrated the use of pointers to allow changes made to a variable within a function to have an effect outside the function, and showed how this could be used to allow a function to, in effect, return more than one variable. An alternative to using pointers is to use *reference variables*: these are variables that are used inside a function that are a different name for the same variable as that sent to a function. When using reference variables any changes inside the function will have an effect outside the function. These are much easier to use than pointers: all that has to be done is the inclusion of the symbol & before the variable name in the declaration of the function and the prototype—this indicates that the variable is a reference variable. It is actually the case that references behave like pointers behind the scenes, but without the programmer having to convert to an address with & on the function call (as in Listing 5.1) and without having to de-reference inside the function—they provide a layer of syntactic sugar to ease the programmer's burden. We now modify the example code in Listing 5.1 that wrote a function that calculated the real and imaginary parts of a complex number given in polar form to use references instead of pointers.

```cpp
#include <iostream>
#include <cmath>

void CalculateRealAndImaginary(double r, double theta,
                               double& real,
                               double& imaginary);

int main(int argc, char* argv[])
{
    double r = 3.4;
    double theta = 1.23;
    double x, y;
    CalculateRealAndImaginary(r, theta, x, y);
    std::cout << "Real part = " << x << "\n";
    std::cout << "Imaginary part = " << y << "\n";

    return 0;
}

void CalculateRealAndImaginary(double r, double theta,
                               double& real,
                               double& imaginary)
{
    real = r*cos(theta);
    imaginary = r*sin(theta);
}
```

5.4 Default Values for Function Arguments

If we are writing a function to implement an iterative technique, such as the Newton–Raphson technique for finding a root of a nonlinear equation, we will usually be content if the solution is accurate to within a tolerance of, say, 10^{-6}. Only on very rare occasions would we want to change this tolerance. We might also want to restrict the number of function evaluations: the Newton–Raphson iteration will probably be implemented using a `while` loop, and numerical rounding errors may prevent the error being sufficiently small for the iteration to terminate. Under these conditions, we would never exit the `while` loop, and the program that called this function would never terminate. It would therefore be prudent to write a function for implementing the Newton–Raphson technique that sets a default tolerance for the solution, and a default maximum number of iterations. We would then be able to call this function without specifying these default values. However, if we did want to call this function with different values then we would like to be able to do this. This is easily achieved by setting default values in the function prototype. This is demonstrated below in a program that uses the Newton–Raphson technique for calculating the cube root of a given number K through solving the nonlinear equation $f(x) = x^3 - K = 0$. Using a given initial guess x_0, the Newton–Raphson method results in the iteration

$$ x_n = x_{n-1} - \frac{x_{n-1}^3 - K}{3x_{n-1}^2}, \quad n = 1, 2, 3, \ldots . $$

By setting default values for the tolerance and maximum number of function iterations we may call the function using one of: (i) the default values of these parameters; (ii) specifying the tolerance (the first optional parameter in the function prototype) and using the default maximum number of function iterations; and (iii) specifying both of these parameters. All three of these cases are shown below.

```
1   #include <cmath>
2   #include <iostream>
3
4   void CalculateCubeRoot(double& x, double K,
5                          double tolerance = 1.0e-6,
6                          int maxIterations = 100);
7
8   int main(int argc, char* argv[])
9   {
10      double x = 1.0;
11      double K = 12.0;
12
13      // Calculate cube root using default values
14      CalculateCubeRoot(x, K);
15
16      // Calculate cube root using a tolerance of 0.001 and the
17      // default maximum number of iterations
18      double tolerance = 0.001;
19      x = 1.0; // Restart guess
20      CalculateCubeRoot(x, K, tolerance);
```

```
21
22     // Calculate cube root using a tolerance of 0.001 and a
23     // maximum number of iterations of 50
24     int maxIterations = 50;
25     x = 1.0; // Restart guess
26     CalculateCubeRoot(x, K, tolerance, maxIterations);
27
28     return 0;
29   }
30
31   void CalculateCubeRoot(double& x, double K,
32                          double tolerance, int maxIterations)
33   {
34     int iterations = 0;
35     double residual = x*x*x-K;
36     while ((fabs(residual) > tolerance) &&
37            (iterations < maxIterations))
38     {
39       x = x-(x*x*x-K)/(3.0*x*x);
40       residual = x*x*x-K;
41       iterations++;
42     }
43   }
```

5.5 Function Overloading

Suppose we want to write one function to multiply a vector by a scalar, and another function to multiply a matrix by a scalar. It would seem natural to call both these functions Multiply. This is allowed in C++: we write different function prototypes and functions for both of these operations: the compiler then chooses the correct function based on the input arguments. This is demonstrated in the code below, and is known as *function overloading*.

```
1   #include <iostream>
2
3   void Multiply(double scalar, double* u, double* v, int n);
4
5   void Multiply(double scalar, double** A, double** B, int n);
6
7   int main(int argc, char* argv[])
8   {
9     int n = 2;
10    double* u = new double [n];
11    double* v = new double [n];
12    double** A = new double* [n];
13    double** B = new double* [n];
```

```
14      for (int i=0; i<n; i++)
15      {
16         A[i] = new double [n];
17         B[i] = new double [n];
18      }
19
20      u[0] = -8.7;   u[1] = 3.2;
21      A[0][0] = 2.3;  A[0][1] = -7.6;
22      A[1][0] = 1.3;  A[1][1] = 45.3;
23      double s = 2.3, t = 4.8;
24
25      // vector multiplication
26      Multiply(s, u, v, n);
27
28      // matrix multiplication
29      Multiply(t, A, B, n);
30
31      delete[] u;
32      delete[] v;
33      for (int i=0; i<n; i++)
34      {
35         delete[] A[i];
36         delete[] B[i];
37      }
38      delete[] A;
39      delete[] B;
40
41      return 0;
42   }
43
44   void Multiply(double scalar, double* u, double* v, int n)
45   {
46      // v = scalar*u (scalar by vector)
47      for (int i=0; i<n; i++)
48      {
49         v[i] = scalar*u[i];
50      }
51   }
52
53   void Multiply(double scalar, double** A, double** B, int n)
54   {
55      // B = scalar*A (scalar by matrix)
56      for (int i=0; i<n; i++)
57      {
58         for (int j=0; j<n; j++)
59         {
60            B[i][j] = scalar*A[i][j];
61         }
62      }
63   }
```

Note that we can overload functions based only on the number and type of the *arguments* and not on the return type. This means that we could not have vector multiply function `bool Multiply(double scalar, double* u, double* v, int n)` alongside the version which has a `void` return type. This is because the compiler can infer the correct version of an overloaded function from the types of its arguments from the context in which it is used. This is not the case with the return type, where you may want to call a function which returns something, but then to cast its output to another return type, or ignore its output completely.

5.6 Declaring Functions Without Prototypes

It is good practice to give the function signature prototypes before you write the implementation. This is so that the function `main`, or any other function will recognise the name and argument types of the new function. However, it is possible to skip the writing of the function prototype by writing the function implementation before its first use, as is shown in the code below.

```
1  #include <iostream>
2
3  double Square(double x)
4  {
5     return x*x;
6  }
7
8  int main(int argc, char* argv[])
9  {
10    std::cout << "Square of 2 = " << Square(2) << "\n";
11    return 0;
12 }
```

If prototypes are not given, then the function implementations must be ordered in such a way that each implementation is seen by the compiler before its first use. Note that if two functions are mutually recursive, that is, both functions call the other function, then it will not be possible to order the functions in this way—and so prototypes must be declared in this case.

5.7 Function Pointers

Suppose we want to write a function to implement the solution of the nonlinear equation $f(x) = 0$ using the Newton–Raphson technique, where f is a user-specified function. We may want to call this function for solving nonlinear equations more than once during the execution of a given program, and for different user-specified nonlinear functions. To achieve this, we need to specify the appropriate nonlinear function each time the function is called. This may be done, as demonstrated in the code below, using *function pointers*.

In the code below, we specify two functions myFunction and myOther-Function. In line 8, we declare a *function pointer* *p_function. This declaration specifies that the function that this pointer refers to must: (i) accept one (and only one) input argument which is a double precision floating point variable; and (ii) return one double precision floating point variable. In line 10, we specify that p_function points at the function myFunction: calling the function p_function in line 11 then has an identical effect to calling myFunction. In lines 13 and 14, we demonstrate how to use p_function to subsequently call the function myOtherFunction.

```
1   #include <iostream>
2
3   double myFunction(double x);
4   double myOtherFunction(double x);
5
6   int main(int argc, char* argv[])
7   {
8      double (*p_function)(double x);
9
10     p_function = &myFunction;
11     std::cout << p_function(2.0) << "\n";
12
13     p_function = &myOtherFunction;
14     std::cout << p_function(2.0) << "\n";
15
16     return 0;
17  }
18
19  double myFunction(double x)
20  {
21     return x*x;
22  }
23
24  double myOtherFunction(double x)
25  {
26     return x*x*x;
27  }
```

The Newton–Raphson method for solving nonlinear equations is defined in Exercise 2.6 in the Exercises at the end of Chap. 2. This is implemented below for two different user-specified functions through the use of function pointers. In lines 5–16, we write a function to implement this algorithm. This function requires specification of: (i) a function pointer to the nonlinear function; (ii) a function pointer to the derivative of the nonlinear function; and (iii) an initial guess to the solution. Note that the function as it stands does not check for divergence, so is unsafe to use in some cases.

In lines 46 and 47, we call the Newton–Raphson solver to solve the equation $\sqrt{x} - 10 = 0$ with initial guess $x = 1$: the nonlinear function Sqrt10, and the derivative of the nonlinear function Sqrt10Prime are given in lines 19–22 and 26–29 of the code. Similarly, in lines 48 and 49 we call the Newton–Raphson solver to solve the equation $x^3 - 10 = 0$ with initial guess $x = 1$: the nonlinear function Cube10, and the derivative of the nonlinear function Cube10Prime are given in lines 32–35 and 39–42 of the code.

```cpp
1   #include <cmath>
2   #include <iostream>
3
4   // Implementation of Newton-Raphson iteration
5   double SolveNewton(double (*pFunc)(double),
6                      double (*pFuncPrime)(double),
7                      double x)
8   {
9       double step;
10      do
11      {
12          step = (*pFunc)(x)/(*pFuncPrime)(x);
13          x -= step;
14      } while (fabs(step) > 1.0e-5);
15      return x;
16  }
17
18  // Function to calculate x that satisfies sqrt(x)=10
19  double Sqrt10(double x)
20  {
21      return sqrt(x) - 10.0;
22  }
23
24  // Derivative of function to calculate x that satisfies
25  // sqrt(x)=10
26  double Sqrt10Prime(double x)
27  {
28      return 1.0/(2.0*sqrt(x));
29  }
30
31  // Function to calculate x that satisfies x*x*x=10
32  double Cube10(double x)
33  {
34      return x*x*x - 10.0;
35  }
36
37  // Derivative of function to calculate x that satisfies
38  // x*x*x=10
39  double Cube10Prime(double x)
```

```
40   {
41       return 3.0*x*x;
42   }
43
44   int main(int argc, char* argv[])
45   {
46       std::cout << "Root sqrt(x)=10, with guess 1.0 is "
47                   << SolveNewton(Sqrt10,Sqrt10Prime,1.0) << "\n";
48       std::cout << "Root x**3=10, with guess 1.0 is "
49                   << SolveNewton(Cube10,Cube10Prime,1.0) << "\n";
50       return 0;
51   }
```

5.8 Recursive Functions

In some applications, we may wish to call a function from within the same function: this is known as *recursion*, and is possible in C++. A simple application of this is the calculation of the factorial of a positive integer n, denoted by fact(n), and written mathematically as *n*!, which is defined by

$$fact(n) = n \times fact(n-1), \quad n > 1,$$
$$fact(n) = 1, \quad n = 1.$$

Code to implement this recursive definition of the factorial function is given below: we simply call the function CalculateFactorial from within the same function as many times as required.

```
1    #include <iostream>
2    #include <cassert>
3
4    int CalculateFactorial(int n);
5
6    int main(int argc, char* argv[])
7    {
8        int n = 7;
9        std::cout << "The factorial of " << n
10                   << " is " << CalculateFactorial(n) << "\n";
11
12       return 0;
13   }
14
```

```
15   int CalculateFactorial(int n)
16   {
17       assert (n > 0);
18       if (n == 1)
19       {
20           return 1;
21       }
22       else
23       {
24           // n>1
25           return n*CalculateFactorial(n-1);
26       }
27   }
```

5.9 Modules

Suppose we want to write a code to allow us to solve linear systems of the form **Ax** = **b**, where **A** is a square, invertible matrix of size n, **b** is a specified vector of size n, and **x** is a vector to be calculated of size n. It would be useful if we could write all the functions required to solve this linear system and then allow these functions to be called through an appropriate function—that is, we want to write a function called SolveLinearSys with the prototype shown below.

```
void SolveLinearSys(double** A, double* x, double* b, int n);
```

The function SolveLinearSys has all the information required to solve the linear system, and any functions required can be called from within this function. This then allows us to solve any suitably defined linear system using just the single line of code shown below.

```
SolveLinearSys(A, x, b, n);
```

The function SolveLinearSys, and all other functions associated with this linear solver, are known as a *module*. In more concrete terms, a module is a collection of functions that performs a given task. Every module has an *interface*. In the example above, this was defined by the prototype of the function SolveLinearSys, and may be thought of as a list of variables that contains: (i) those that must be input to the module; and (ii) those that are output by the module.

Modules are very useful when sharing code. For example, if a colleague has written code for solving linear systems as described above then it would be a very simple task for another colleague to utilise this code. All that is required is an understanding of the interface and what the purpose of the code is: there is no need to understand the mathematical algorithm that determines *how* the linear system has been solved, and the module may be thought of as a "black box".

5.10 Tips: Code Documentation

As you begin to write more programs, there is often a temptation to "just get on with the coding" without paying specific attention to quality. After all "you generally know where you are going and understand the program which you are writing". It is important to bear in mind, though, that your code will not always be as well understood as it is now. You might come back to a given file in three years' time, because you need to correct it or to add some new functionality to it. Alternatively, you may at some stage hand your programs over to someone else who has the job of working out what you were doing.

Our tip in this chapter is that computer programs should be human-readable, as well as machine-readable. Even the smallest portion of code may prove to be opaque unless we include enough commentary to aid the human reader. Take for example the function given below, which calculates the *p*-norm of a vector. Without comments in the code, it would not be obvious what was happening, even though there are only a few lines of code. A hint is given in the name of the function, `CalculateNorm`, but what is it meant to do? What is the significance of the arguments s and p?

```
1   #include <cmath>
2   double CalculateNorm(double* x, int s, int p)
3   {
4       double a = 0.0;
5       for (int i=0; i<s; i++)
6       {
7           double temp = fabs(x[i]);
8           a += pow(temp, p);
9       }
10      return pow(a, 1.0/p);
11  }
```

In the code segment below, we give a description of the function immediately before its definition. This description gives, in line 3, a means of mapping the mathematics of the function to its implementation. The rest of the description gives an alternate place to find more information about the *p*-norm (lines 4–6) and an explanation of some of the arguments as necessary. In the body of the function, the loop has been commented to describe what its *functional purpose* is: it is about computing a sum over the elements of the vector. Finally, the return value is commented with a few words of explanation.

```
1   #include <cmath>
2   //  Function to calculate the p-norm of a vector:
3   //      = [ Sum_i ( |x_i|**p ) ] **(1/p)
4   //  See "An Introduction to Numerical Analysis" by
5   //  Endre Suli and David Mayers, page 60, for definition
6   //  of the p-norm of a vector
7   //  x is a pointer to the vector which is of size vecSize
8
9   double CalculateNorm(double* x, int vecSize, int p)
10  {
11      double sum = 0.0;
12      //Loop over elems x_i of x, incrementing sum by |x_i|**p
13      for (int i=0; i<vecSize; i++)
14      {
15          double temp = fabs(x[i]);
16          sum += pow(temp, p);
17      }
18      //Return p-th root of sum
19      return pow(sum, 1.0/p);
20  }
```

Note that documenting code is sometimes more of an art than a science. There is a balance to be struck concerning the right level of documentation. Too many comments can make the program less readable rather than more readable. Our tip here is that you should describe what part of the problem the code is solving and, perhaps, *how* it is solving that problem. Do not be tempted to describe the code in overmuch detail. For example, the comment on the loop in line 12 of code above could have read

```
12      //  Loop over values of i going from 0 to vecSize-1
```

While this comment is accurate (describing the range of the loop variable vecSize), it does nothing to aid a programmer in their understanding of the code.

The formatting of the code documentation can also help readability. A simple tip is that using empty lines to break code and comments into sections can make the code look more readable. If you want to emphasise something you can simulate underlining with hyphens or underscores, for example,

```
4       //  Very important comment
5       //  ----------------------
```

Alternatively, you can emphasise something by putting it in a box:

```
2   /***************************************************
3       ***************************************************
4       **              CalculateNorm(...)              **
5       **                                              **
6       **    Function to calculate p-norm of vector    **
7       ***************************************************
8       ***************************************************/
```

5.11   Exercises

In all exercises, we suggest that you use dynamic allocation of memory for vectors and matrices as described in Sect. 4.2. Be sure that you are correctly de-allocating memory when using dynamic allocation of memory, as explained in the exercises at the end of Chap. 4.

5.1 Write code that sends the *address* of an integer to a function that prints out the *value* of the integer.

5.2 Write code that sends the address of an integer to a function that changes the value of the integer.

5.3 Write a function that swaps the values of two double precision floating point numbers, so that these changes are visible in the code that has called this function.
1. Write this function using pointers.
2. Write this function using references.

5.4 Write a function that can be used to calculate the mean and standard deviation of an array of double precision floating point numbers. Note that the standard deviation σ of a collection of numbers x_j, $j = 1, 2, \ldots, N$ is given by

$$\sigma = \sqrt{\frac{\sum_{j=1}^{N}(x_j - \bar{x})^2}{N - 1}}$$

where $\bar{x}$ is the mean of the numbers.

5.5 Write a function `Multiply` that may be used to multiply two matrices given the matrices and the size of both matrices. Use assertions to verify that the matrices are of suitable sizes to be multiplied.

5.6 Overload the function `Multiply` written in the previous exercise so that it may be used to multiply:

1. a vector and a matrix of given sizes;
2. a matrix and a vector of given sizes;
3. a scalar and a matrix of a given size; and
4. a matrix of a given size and a scalar.

5.7 The p-norm of a vector $\mathbf{v}$ of length n is given by

$$\|\mathbf{v}\|_p = \left(\sum_{i=1}^n |v_i|^p\right)^{1/p}$$

where p is a positive integer. Extend the code in Sect. 5.10 to calculate the p-norm of a given vector, where p takes the default value 2.

5.8 The determinant of a square matrix may be defined recursively: see Sect. A.1.3. Write a recursive function that may be used to calculate the determinant of a square matrix of a given size. Check the accuracy of your code by comparison with the known formulae for square matrices of size 2 and 3:

$$\det\begin{pmatrix} A_{00} & A_{01} \\ A_{10} & A_{11} \end{pmatrix} = A_{00}A_{11} - A_{01}A_{10},$$

$$\det\begin{pmatrix} A_{00} & A_{01} & A_{02} \\ A_{10} & A_{11} & A_{12} \\ A_{20} & A_{21} & A_{22} \end{pmatrix} = A_{00}(A_{11}A_{22} - A_{12}A_{21}) - A_{01}(A_{10}A_{22} - A_{12}A_{20})$$

$$+ A_{02}(A_{10}A_{21} - A_{11}A_{20}).$$

5.9 Write a module for solving the 3×3 linear system $\mathbf{Au} = \mathbf{b}$ where $\mathbf{A}$ is nonsingular.

5.10 Write a module for solving the $n \times n$ linear system $\mathbf{Au} = \mathbf{b}$ using Gaussian elimination with pivoting, where $\mathbf{A}$ is nonsingular. See Sect. A.2.1.3 for details of this algorithm.

An Introduction to Classes

One of the key features of the C++ programming language is that it is *object-oriented*. Up until now we have largely ignored this feature, making only passing reference to it in earlier chapters. For the remainder of this book, we focus on object-orientation, allowing readers to utilise this feature in their C++ programs.

6.1 The *Raison d'Être* for Classes

At the end of Chap. 5 we introduced the concept of a module. We explained that modules are useful for code reuse, and therefore allow rapid code development for programs that require the functionality provided by the module, even if the programmer has no understanding of the operations that a module performs. This may be highlighted by using the example of a module for solving linear systems that was introduced in Chap. 5. Three advantages of having this module available are given below.

- Linear algebra lies at the heart of numerical analysis, and so numerical analysts use linear solvers in many programs that they write. A module allows them to reuse this code rather than write new functionality for solving linear systems each time they write a new program.
- There are many different linear algebra techniques for solving linear systems. It is possible to include many different techniques in a module, and to specify which technique is to be used as part of the interface to the module.
- Other scientists with little mathematical expertise may have to write programs which require the solution of a linear system. A module allows them to do so without learning the mathematical techniques that underpin linear algebra algorithms.

© Springer International Publishing AG, part of Springer Nature 2017
J. Pitt-Francis and J. Whiteley, *Guide to Scientific Computing in C++*, Undergraduate Topics in Computer Science, https://doi.org/10.1007/978-3-319-73132-2_6

Modules are clearly very useful when writing scientific computing programs. But, as we now explain, the use of modules may cause problems.

6.1.1   Problems That May Arise When Using Modules

Suppose that the linear solver that we discussed in the previous section has been written so that the solution of this linear system is calculated using the GMRES algorithm.[1] This technique for solving linear systems requires several instances of a calculation of the scalar product between two vectors. Implementation of this technique would, therefore, probably include a function being written to calculate the scalar product of two vectors of a given length. Use of this function would not be restricted to users of the module for solving linear systems: another part of the code may use this function to calculate, for example, the normal derivative of a function of two or more variables. Suppose whoever was using the scalar product function to calculate a normal derivative decided to change the inputs to the scalar product function. This would inadvertently cause the linear solver to stop functioning correctly. The linear solver module could then *not* be treated as a "black box".

Another drawback of using standard modules is the way in which data is stored. There is only ever one copy of a particular module and one copy of any data associated with it. If that data is changed for the module to fulfil a particular purpose, then it will be changed for all future uses. Consider a linear solver which has had its functionality extended so that it is able to deal with singular matrices. Such a linear solver will need to have access to the *null space* (or kernel) of the singular matrix or matrices in question. Suppose we use the extended linear solver to solve a singular linear system. The linear system will then solve the singular system subject to knowing and storing the null space of this system. If we were to subsequently use the module to solve another nonsingular linear system, we would have to remember to specify the null space as being empty or the linear solver would attempt to find the solution of the nonsingular system subject to the previously specified null space.

In the next section, we explain how *classes* allow us to write code including all the features of modules, but without the drawbacks identified above.

6.1.2   Abstraction, Encapsulation and Modularity Properties of Classes

The shortcomings of modules, described in the previous section using the example of a module for solving a linear system, could be overcome if we could write a "module" that:

---

[1]The Generalised Minimal RESidual technique—commonly known as GMRES—is an iterative technique for solving linear systems. See, for example, Trefethen and Bau [4] for more details.

1. contains all the functions needed to solve the system;
2. does not allow these functions to be accessed by any other part of the program except through the interface;
3. can not itself access any other part of the program; and
4. also contains all the data needed to solve the system.

This is possible through the use of *classes*, and the specifications described above—that is, the compartmentalisation of all of the resources needed—are known as the *encapsulation* feature of classes. The variables/data and functions associated with a class are known as *class members*, and the functions more specifically as *methods*. We are now in a position to describe some of the technical terms from Sect. 1.1.1.

Classes allow *modularity*, which includes placing similar functionality in a few files. Classes allow us to go further than this: access controls allow us to control which resources are available outside of the class, and which are hidden from users. Hiding parts of the code may—at first sight—seem to have the undesirable effect of preventing a user from accessing the full functionality of the software. As we shall see later in this chapter, this is certainly not the consequence: it actually has the more desirable effect of preventing users from inadvertently corrupting data. Furthermore, combining functionality in this way allows us to associate data with the functionality.

The concept of *abstraction* is that the particulars of an idea should not be important. Classes allow us to hide the irrelevant details of functionality from users who need not know about them. For example, a reader of this book does not need to know *how* a compiler translates a C++ code into a machine readable executable file, but only how to *instruct* the compiler to perform this task. Abstraction allows emphasis to be placed on the qualities or properties that characterise the objects in how they act and the type of information that they carry.

A further property of classes is *inheritance* which allows easy code reuse, extensibility and polymorphism. Inheritance will be discussed in Chap. 7.

6.2 A First Example Simple Class: A Class of Books

The first simple class that we develop is a class of books.

6.2.1 Basic Features of Classes

Each book has the following attributes:
- an author;
- a title;
- a format;
- a price;
- a year of publication; and
- a publisher.

These attributes can be associated with each instance of a book by first saving the file below as `Book.hpp`. As explained earlier, these attributes are known as *class members*.

```
1  #include <string>
2
3  class Book
4  {
5  public:
6      std::string author, title, publisher, format;
7      int price; //Given in pence
8      int yearOfPublication;
9  }; //Note that the class ends with ;
```

The file above is known as the *header file* associated with the class: the extension .hpp indicates that this file is a header file associated with a C++ program. At this stage, it is sufficient to know that the word `public` that is used in line 5 of this file allows us to access all variables associated with the class. We will give more precise details on what are known as *access privileges* later in this chapter. Note the semi-colon that is required after the closing curly bracket at the end of this file. A common mistake made by novice programmers is to miss this semi-colon at the end of the class definition.

The class of books may then be used as shown in the code below. Note that when header files that we have written are included the names of these files are enclosed within quotation marks, in contrast to the system header files such as `iostream`, `fstream` and `cmath` that we have used earlier. The compiler does not distinguish between included files with quotation marks and those with angle brackets, but a common coding convention encourages programmers to use quotation marks and angle brackets to make the distinction between local include files and those from external libraries, respectively.

Listing 6.1 Using the class Book

```
1  #include <iostream>
2  #include "Book.hpp"
3
4  int main(int argc, char* argv[])
5  {
6      Book my_favourite_book;
7
8      my_favourite_book.author = "Lewis Carroll";
9      my_favourite_book.title =
10                 "Alice's adventures in Wonderland";
11     my_favourite_book.publisher = "Macmillan";
12     my_favourite_book.price = 199;
13     my_favourite_book.format = "hardback";
14     my_favourite_book.yearOfPublication = 1865;
```

```
15
16        std::cout << "Year of publication of "
17                  << my_favourite_book.title << " is "
18                  << my_favourite_book.yearOfPublication << "\n";
19    }
```

The class of books written here allows us to associate data with each instance of the class. As such, we can think of this class as allowing us to define a new data type and line 6 of the code above as declaring an instance of that class, in this case called my_favourite_book. The class members can all be accessed as shown in lines 8–18 of the code above—that is, the string my_favourite_book.author is the class member author associated with the instance of the class called my_favourite_book.

6.2.2   Header Files

It doesn't matter if we include header files such as iostream, string, etc. more than once. But we should be very careful not to include files such as Book.hpp in the form that it was written in the previous section more than once, as this can cause problems. We will see later on in this book when we are working with several different classes that it is easy to inadvertently include header files more than once. To avoid this code being included twice, we adapt it so that the header file for a class called ExampleClass is of the form shown below.

Initially EXAMPLECLASSHEADERDEF will *not* be defined. The "ifndef" in line 1 is a contraction of **if n**ot **def**ined. The first line of code below therefore instructs the computer to include the code between here and the #endif (line 18 of the code) *only if* the *macro* EXAMPLECLASSHEADERDEF is *not* defined. The first time this code is included this macro will not be defined, and so all of the code in the listing below will be read. Note that when this code is included, the first task that is performed is to define the macro EXAMPLECLASSHEADERDEF (line 7 of the code). As EXAMPLECLASSHEADERDEF is now defined, if this code were to be included a second time all code between the #ifndef EXAMPLECLASSHEADER-DEF statement (line 1) and #endif (line 18) will now *not* be included. We therefore see that the #ifndef, #define and #endif statements may be used to ensure that the contents of a header file are not included more than once.

```
1    #ifndef EXAMPLECLASSHEADERDEF    // only if macro
2                                     // EXAMPLECLASSHEADERDEF not
3                                     // defined execute lines of
4                                     // code until #endif
5                                     // statement
6
7    #define EXAMPLECLASSHEADERDEF    // define the macro
8                                     // EXAMPLECLASSHEADERDEF.
9                                     // Ensures that this code is
```

```
10                                      // only compiled once, no
11                                      // matter how many times it
12                                      // is included
13  class ExampleClass
14  {
15     lines of code  // body of header file
16  };
17
18  #endif // need one of these for every #ifndef statement
```

6.2.3 Setting and Accessing Variables

In the class of books we developed in the previous section, all class members were variables, such as strings, double precision floating point numbers, or integers. Classes are, however, much more powerful than this: we will now show how functions may also be defined as class members, known as class *methods*.

Suppose we want to check that the year of publication of an instance of the class Book always takes a valid year. Assuming that no book in our catalogue was published before the invention of the printing press, and has already been published or will be in the near future, then we may write a function known as a *member method*, called SetYearOfPublication, that allows us to set this variable and check that the integer value for year of publication falls within a sensible range (after the invention of the printing press and not too far in the future). As we are writing a method that allows us to check that a valid year of publication is assigned, it seems sensible to force the user of the class to use this method to set this variable. This may be implemented by setting the member yearOfPublication to be a *private variable*. Private variables may only be accessed by other class members: making yearOfPublication a private variable therefore prevents us from accessing this variable through code such as line 14 in Listing 6.1. However, it can be set through the member method SetYearOfPublication, which we will make a public member of this class. Access privileges—that is, the use of public and private members—will be discussed more fully in Sect. 6.2.5.

Now that we have made yearOfPublication a private member, we cannot directly access this member from outside the class. We therefore need to write a public method that allows us to access this member—this class member will be called GetYearOfPublication. We are also going to slightly modify the name yearOfPublication to mYearOfPublication, where the prefix "m"—with the "m" pertaining to "my"—reminds us that this variable is private to the class. We now present code that implements this discussion. First we need a new header file Book.hpp, given below.

In the code below, all members that follow public and precede private (lines 9–12) may be accessed from outside the class. As mYearOfPublication comes after private it is only accessible to class members. We will discuss access privileges more fully in Sect. 6.2.5. Note the methods declared in lines 11 and 12 of this code. We have specified that the method SetYearOfPublication accepts an integer argument and returns no value, that is, it is a void function. The method GetYearOfPublication returns an integer, but does not require any input arguments as it can access all class members including mYearOfPublication. The keyword const after the declaration of this method is a signal to the compiler that we want to ensure that the instance of the class will remain constant throughout the execution of the method. That is, the method GetYearOfPublication should have changed nothing inside the class. We now need to tell the computer what these methods do. This is given in the code in Listing 6.3, which should be saved as Book.cpp. We have used an assert statement to check that the year of publication does fall within a sensible period when it is set. Note that the header file required for assert statements should be included in this file.

Listing 6.2 The file Book.hpp

```
 1  #ifndef BOOKHEADERDEF
 2  #define BOOKHEADERDEF
 3
 4  #include <string>
 5
 6  class Book
 7  {
 8  public:
 9      std::string author, title, publisher, format;
10      int price; //Given in pence
11      void SetYearOfPublication(int year);
12      int GetYearOfPublication() const;
13  private:
14      int mYearOfPublication;
15  };
16
17  #endif
```

Listing 6.3 The file Book.cpp

```
 1  #include <cassert>
 2  #include "Book.hpp"
 3
 4  void Book::SetYearOfPublication(int year)
 5  {
 6      assert ((year > 1440) && (year < 2020));
 7      mYearOfPublication = year;
 8  }
 9
10  int Book::GetYearOfPublication() const
11  {
12      return mYearOfPublication;
13  }
```

In the code above, line 4 requires more explanation. In common with functions introduced in Chap. 5, the void at the start of this line indicates that this method does not return any variable. The remainder of this line indicates that this method: (i) is associated with a class called Book; (ii) is called SetYearOfPublication; and (iii) requires one integer input argument which will be termed year. Inside this method we first check that the input year is appropriate through an assertion, before allocating it to the mYearOfPublication of a book. The method GetYearOfPublication, which is written in lines 10–13, allows us to access the variable mYearOfPublication from outside the class, without allowing us to change this value to what may be an incorrect value.

Code that uses this updated class is given below, and should be saved as Use-BookClass.cpp. Using access privileges to ensure that variables may only be set through a class member that provides a check on the accuracy of data is very good programming practice, and should be used whenever possible.

Listing 6.4   The file UseBookClass.cpp

```
1   #include <iostream>
2   #include "Book.hpp"
3
4   int main(int argc, char* argv[])
5   {
6       Book promotion_book;
7
8       promotion_book.author = "Iris Murdoch";
9       promotion_book.title = "The sea, the sea";
10      promotion_book.publisher = "Chatto & Windus";
11      promotion_book.price = 299;
12      promotion_book.format = "hardback";
13      promotion_book.SetYearOfPublication(1978);
14
15      std::cout << "Year of publication of "
16              << promotion_book.title << " is "
17              << promotion_book.GetYearOfPublication()
18              << "\n";
19  }
```

Note that in line 17 of the code above we need to acknowledge that the class member GetYearOfPublication is a function or method by including empty brackets after using this class method, even though no input arguments are required.

The files Book.hpp and Book.cpp together form valid C++ code for a class of books. The code in Listing 6.4 above is a valid C++ use of this class. So far in this book we have only needed to compile one file. Now, however, we need to think a bit more about how to compile the multiple files that arise from using classes.

6.2.4   Compiling Multiple Files

In Sect. 1.3.3 we compiled a single C++ file into an executable program using the single compilation step below.

```
g++ -Wall -o HelloWorld HelloWorld.cpp
```

What really happens in this process is that the C++ file is first compiled to another file called `HelloWorld.o`, and known as an *object file*, which is a machine-readable file. In a second step, the object file is compiled into the executable file and the intermediate object file is deleted. What we are actually doing when using the compilation command above is to combine the two compilation steps given below.

```
g++ -Wall -c HelloWorld.cpp
g++ -Wall -o HelloWorld HelloWorld.o
```

The first of these commands creates an object file called `HelloWorld.o` from the C++ file `HelloWorld.cpp` through the use of the `-c` compiler flag. The second command creates an executable file `HelloWorld` from the object file `HelloWorld.o`. Up until this point, we have used a one line compilation command, allowing us to completely ignore the existence of object files. When compiling multiple files we do, however, need to be aware of the existence of these files.

Before we can compile the file `UseBookClass.cpp` in Listing 6.4, we first need to compile the `Book` class to create an object file `Book.o` associated with this class. This is done, as above, by using the `-c` option when compiling:

```
g++ -Wall -O -c Book.cpp
```

This produces an object file `Book.o`. We can now compile `UseBookClass.cpp` into an object file and then *link* the two object files to make an executable. The two compilation commands are now

```
g++ -Wall -O -c UseBookClass.cpp
g++ -Wall -lm -O -o UseBookClass UseBookClass.o Book.o
```

As in the above "HelloWorld" example, it is possible to skip one step in the compilation process so that we do not have to explicitly produce the intermediate file `UseBookClass.o`.

```
g++ -Wall -lm -O -o UseBookClass UseBookClass.cpp Book.o
```

The code may be run as before by typing

```
./UseBookClass
```

at the command line.

6.2.4.1 Using Makefiles to Compile Multiple Files

Suppose we have code that uses several classes stored in several files. We would rather not compile *all* of these classes separately every time one file is modified slightly. This may be avoided by the use of a Makefile—using this approach only the necessary compilation is carried out. The following is a Makefile for code UseClasses.cpp that uses two classes, Class1 and Class2.

```
1  Class1.o : Class1.cpp Class1.hpp
2          g++ -c -O Class1.cpp
3
4  Class2.o : Class2.cpp Class2.hpp
5          g++ -c -O Class2.cpp
6
7  UseClasses.o : UseClasses.cpp Class1.hpp Class2.hpp
8          g++ -c -O UseClasses.cpp
9
10  UseClasses : Class1.o Class2.o UseClasses.o
11          g++ -O -o UseClasses Class1.o Class2.o UseClasses.o
```

If the file above is saved as Makefile, then to generate an up-to-date executable file UseClasses we simply type "make UseClasses" at the command line.

Using this approach only the necessary compilation will be carried out. Line 10 of this Makefile tells the compiler that the executable file UseClasses requires three files: Class1.o, Class2.o and UseClasses.o. Line 11 gives the rule for compiling the executable file from its dependencies. Line 1 tells the compiler that the file Class1.o depends on the two files Class1.cpp and Class1.hpp. Only if one or both of these files have been changed since the last time this class has been compiled will this class be recompiled using the rule given on Line 2. Similar remarks hold for the class Class2. Note that in line 7, the recompilation of UseClasses.o depends not only on the relevant C++ file, but also on the classes' header files—so that a change in either class interface will result in a recompilation of the file which uses its functionality. Finally, having worked through all the steps described, a new executable UseClasses will be created only if one or more of the files listed on line 10 have changed as a consequence of this compilation process.

The compilation procedure is illustrated in Fig. 6.1. In this figure, the thin lines with arrows represent some of the code dependencies described above that are encapsulated within the Makefile. Many of the integrated development environments described in Sect. 1.3.1 will automatically generate Makefiles.

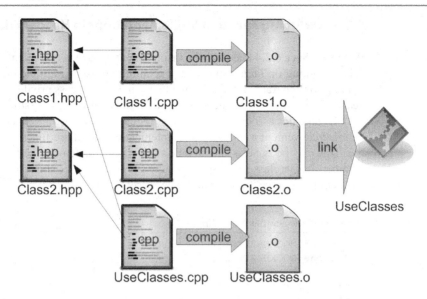

Fig. 6.1 The compilation process

6.2.5   Access Privileges

In Sect. 6.2.3, we briefly discussed access to class members. There are three degrees of access to class members:

- private—these class members are only accessible to other class members, unless friend (which will be introduced in Sect. 6.3) is used;
- public—these class members are accessible to everyone;
- protected—these class members are accessible to other class members, to derived classes (which will be introduced in Chap. 7), and to friends.

The reserved keywords private, public and protected may be used as often as desired, with the default being private. For example in the class below, member1 and member3 are private members, member2 and member4 are public members, and member5 is a protected member.

```
1   #include <string>
2   class ExampleClass
3   {
4       double member1;
5   public:
6       std::string member2;
7   private:
8       int member3;
9   public:
10      int member4;
11  protected:
12      double member5;
13  };
```

6.2.6 Including Function Implementations in Header Files

We saw in Sect. 6.2.4 that it can be inconvenient to have to compile multiple classes. When working on large projects that require the use of multiple classes it can be difficult to keep track of the class members and their access privileges (stored in the header file) and the implementations of the member functions (stored in the .cpp file). If functions associated with a class require only a few lines of code then it may be more convenient to include the implementation of these functions in the header file. This may be done as shown below, where we implement the functions that are members of our class Book in the header file for this class, thus combining the files in Listings 6.2 and 6.3 into a single file Book.hpp.

Listing 6.5 The new file Book.hpp

```
1   #ifndef BOOKHEADERDEF
2   #define BOOKHEADERDEF
3
4   #include <string>
5   #include <cassert>
6
7   class Book
8   {
9   public:
10      std::string author, title, publisher, format;
11      int price; //Given in pence
12      void SetYearOfPublication(int year)
13      {
14          assert ((year > 1440) && (year < 2020));
15          mYearOfPublication = year;
16      }
17      int GetYearOfPublication() const
18      {
19          return mYearOfPublication;
20      }
21   private:
22      int mYearOfPublication;
23   };
24
25   #endif
```

6.2.7 Constructors and Destructors

Each time an object of the class Book is created the program calls a function that allocates space in memory for all the variables used. This function is called a *default constructor* and is automatically generated. This default constructor can be overridden if desired—for example we may wish to set all the string variables in our class

of books to "unspecified" so that it will be clear when accessing this object that
these strings have not yet been properly assigned. An appropriate header file for this
class is shown below. Note that when overriding the default constructor this function
has the same name as the class, takes no arguments, has no return type and must be
a public member of the class.

```cpp
#ifndef BOOKHEADERDEF
#define BOOKHEADERDEF

#include <string>

class Book
{
public:
    Book();
    std::string author, title, publisher, format;
    int price;   //Given in pence
    void SetYearOfPublication(int year);
    int GetYearOfPublication() const;
private:
    int mYearOfPublication;
};

#endif
```

The methods associated with this class are given in the file below.

```cpp
#include "Book.hpp"
#include <cassert>

//This overrides the default constructor
Book::Book()
{
    author = "unspecified";
    title = "unspecified";
    publisher = "unspecified";
    format = "unspecified";
}

void Book::SetYearOfPublication(int year)
{
    assert ((year > 1440) && (year < 2020));
    mYearOfPublication = year;
}

int Book::GetYearOfPublication() const
{
    return mYearOfPublication;
}
```

The code below demonstrates how to use the overridden default constructor.

```cpp
#include <iostream>
#include "Book.hpp"

int main(int argc, char* argv[])
{
    Book my_book;
    std::cout << "The author is " << my_book.author << "\n";

    return 0;
}
```

The code above will print "The author is unspecified".

We will see in Chap. 10 that, if any memory management such as allocating memory dynamically is required by a class, then it is *essential* to change the behaviour of the automatically generated default constructor: if not, the default constructor will not allocate any memory. We can change the behaviour of the automatically generated default constructor either by overriding it with a default constructor of our own (as in the example of Book, above) or by providing some other constructor (which we will discuss shortly). This is because the automatically generated default constructor is *only available* if no other constructors have been provided by the programmer.

Another constructor that is automatically generated is a *copy constructor*. This constructor requires as input another instance of the class, and creates a copy of this instance of the class. In common with default constructors, copy constructors may also be overridden. Note that the argument to a copy constructor has to be a *reference* to another instance of the class, rather than that object itself. This is because, by default, all method arguments are called by copy. Were we to miss the fact that this constructor takes a reference argument, then we would need to use a copy constructor in the call—the very machinery that we are defining here. It is also a good idea to declare the argument to a copy constructor as const which is an instruction to the compiler to ensure that the object argument otherBook to the copy constructor in the code in Listing 6.6 will remain constant during this operation. That is, the constructor will have no hidden side-effects on the instance of the class that it is copying.

Furthermore, in addition to the default and copy constructors, we may write our own customised constructor that takes any inputs that we feel are appropriate, and we may write as many of these constructors as we like. For example, we may want to specify a book's title when creating an object. We now demonstrate how to write a constructor such as this, and how to override a copy constructor. First, we need an appropriate header file: one is shown in Listing 6.6. Line 10 of this header file declares an overridden copy constructor, and line 11 explains that there will be a constructor that accepts a string as input. As we have provided a constructor ourselves the automatically generated default constructor is *not* available: we may, however, supply a default constructor ourselves.

The methods associated with this class are given in the file in Listing 6.7. Lines 14–22 are the overridden copy constructor, where all class members are set to be the same as the instance of the class that we wish to copy. Lines 25–28 represent the specialised constructor that sets the title of the book to a specified string.

The code in Listing 6.8 first creates an instance of the class Book, called good_read, and sets the class members associated with good_read. Line 15 demonstrates how to use the overridden copy constructor to create another instance of the class Book, called another_book, that is initialised with class members taking identical values to those of good_read. Line 17 uses the constructor that sets the title when the instance of the class is declared: an instance of the class called an_extra_book is declared, with title set to "The Magician's nephew".

Destructors are also automatically written, and free memory allocated for an object when it goes out of scope. We will see later when writing classes of vectors and matrices that there are situations—specifically where the constructor has performed dynamic allocation of memory—where the automatically generated destructor *should* be overridden. This allows us to adhere to the tip introduced in Sect. 4.3.3, which advised programmers to ensure that any line of code where memory is dynamically allocated using new has a corresponding line where the memory is freed up using delete.

Listing 6.6 The file Book.hpp

```
 1  #ifndef BOOKHEADERDEF
 2  #define BOOKHEADERDEF
 3
 4  #include <string>
 5
 6  class Book
 7  {
 8  public:
 9      Book();
10      Book(const Book& otherBook);
11      Book(std::string bookTitle);
12      std::string author, title, publisher, format;
13      int price; //Given in pence
14      void SetYearOfPublication(int year);
15      int GetYearOfPublication() const;
16  private:
17      int mYearOfPublication;
18  };
19
20  #endif
```

Listing 6.7 The file `Book.cpp`

```cpp
#include "Book.hpp"
#include <cassert>

//Overridden default constructor
Book::Book()
{
    author = "unspecified";
    title = "unspecified";
    publisher = "unspecified";
    format = "unspecified";
}

//Overridden copy constructor (mimics system version)
Book::Book(const Book& otherBook)
{
    author = otherBook.author;
    title = otherBook.title;
    publisher = otherBook.publisher;
    format = otherBook.format;
    price = otherBook.price;
    mYearOfPublication = otherBook.GetYearOfPublication();
}

//Specialised constructor
Book::Book(std::string bookTitle)
{
    title = bookTitle;
}

void Book::SetYearOfPublication(int year)
{
    assert ((year > 1440) && (year < 2020));
    mYearOfPublication = year;
}

int Book::GetYearOfPublication() const
{
    return mYearOfPublication;
}
```

Listing 6.8 Example code that uses the Book class

```cpp
#include <iostream>
#include "Book.hpp"

int main(int argc, char* argv[])
{
    Book good_read;
```

```
7
8      good_read.author = "C S Lewis";
9      good_read.title = "The silver chair";
10     good_read.publisher = "Geoffrey Bles";
11     good_read.price = 699;
12     good_read.format = "paperback";
13     good_read.SetYearOfPublication(1953);
14
15     Book another_book(good_read);
16
17     Book an_extra_book("The Magician's nephew");
18
19     return 0;
20  }
```

6.2.8 Pointers to Classes

We may declare a pointer to an instance of a class as we show in the code below. In line 6 of this code we declare a pointer, p_book_i_am_reading, to an instance of the class Book described earlier in this chapter, and allocate memory for this instance through the use of new. In line 8, we use *p_book_i_am_reading to denote the contents of the memory whose address is stored by the pointer. By placing this in brackets, we may access the class members as shown in earlier sections of this chapter. Line 9 is a more convenient way of accessing a class member associated with a pointer to a class in which the forward arrow, ->, means "de-reference and then access the member".

```
1   #include <iostream>
2   #include "Book.hpp"
3
4   int main(int argc, char* argv[])
5   {
6       Book* p_book_i_am_reading = new Book;
7
8       (*p_book_i_am_reading).author = "Philip Pullman";
9       p_book_i_am_reading->title = "Lyra's Oxford";
10
11      delete p_book_i_am_reading;
12  }
```

In the code above, note that we have followed the advice given in Sect. 4.3.3—which we shall repeat many times in this book—to always write a delete statement to match a new statement.

6.3 The `friend` Keyword

When developing a program, we may wish to access private members of a class from outside the class. One way of doing this is to create a new public method that accesses the private member in the same way as we did in Sect. 6.2.3. Another way is to write a free function that is a *friend* of the class: such functions may access all members of the class, including private variables. This is demonstrated in the class that we write below. First, we write the header file.

```
 1  #ifndef EXAMPLECLASSDEF
 2  #define EXAMPLECLASSDEF
 3
 4  class ExampleClass
 5  {
 6  private:
 7     double mMemberVariable1;
 8     double mMemberVariable2;
 9
10  public:
11     ExampleClass(double member1, double member2);
12     double GetMinimum() const;
13     friend double GetMaximum(const ExampleClass& ex_class);
14  };
15
16  #endif
```

The constructor, member function and friend function are then implemented using the code below. Note that as the friend function `GetMaximum` is not a member of the class, we do not include `ExampleClass::` in line 25 of the code as we would do when writing a method that is a member of the class.

```
 1  #include "ExampleClass.hpp"
 2
 3  //Constructor to set private members
 4  ExampleClass::ExampleClass(double member1, double member2)
 5  {
 6     mMemberVariable1 = member1;
 7     mMemberVariable2 = member2;
 8  }
 9
10  //GetMinimum is a member method
11  double ExampleClass::GetMinimum() const
12  {
13     if (mMemberVariable1 < mMemberVariable2)
14     {
15        return mMemberVariable1;
16     }
```

```
17      else
18      {
19          // mMemberVariable1 >= mMemberVariable2
20          return mMemberVariable2;
21      }
22  }
23
24  //GetMaximum is a friend function
25  double GetMaximum(const ExampleClass& eg_class)
26  {
27      if (    eg_class.mMemberVariable1 >
28              eg_class.mMemberVariable2)
29      {
30          return eg_class.mMemberVariable1;
31      }
32      else
33      {
34          // eg_class.Var1 <= eg_class.Var2
35          return eg_class.mMemberVariable2;
36      }
37  }
```

Code that uses the friend function of the class above is shown below.

```
1   #include <iostream>
2   #include "ExampleClass.hpp"
3
4   int main(int argc, char* argv[])
5   {
6       ExampleClass example(2.0, 3.0);
7       std::cout << "Minimum value = " << example.GetMinimum()
8                 << "\n";
9       std::cout << "Maximum value = " << GetMaximum(example)
10                << "\n";
11      return 0;
12  }
```

6.4   A Second Example Class: A Class of Complex Numbers

In the class of books that we have developed, all class members were quite simple, being either variables—such as strings or integers—or straightforward methods. We now develop a class of complex numbers, allowing some more advanced features of classes—such as operator overloading—to be showcased through a scientific computing example. It is worth pointing out, before developing the class, that C++ does already have a complex number type which is based on templates (see Chap. 8). We are developing a complex number class here solely for illustration. If you need to use complex numbers we recommend you use the official C++ class (which we will revisit in Sect. 9.5).

A complex number has a real part and an imaginary part. A class of complex numbers will therefore contain class members that represent both of these quantities. It seems sensible to override the default constructor to set both the real and imaginary part of a complex number to zero in the absence of any specified value. We would also like a constructor to be available that allows us to set the complex number $z = x + iy$, where x and y are double precision floating point variables, using statements of the form shown below.

```
1   double x = 4.0;
2   double y = -3.0;
3   ComplexNumber z(x, y);
```

In addition, we may also include class members that are methods that calculate both the modulus and the argument of this complex number. A further method that may be of use is raising the complex number to a specified power.

6.4.1 Operator Overloading

If we have declared a, b, c and d to be integer variables then we may easily relate these variables through statements such as those below.

```
1   int a, b, c, d;
2   a = b;
3   c = -a;
4   d = a + b;
```

We would also like to write statements such as these if a, b, c and d were complex numbers rather than integers. Before we can do this, we need to define: (i) what the assignment operator (equals) means for complex numbers; (ii) what the *unary*[2] minus operator means—i.e. what is meant by the expression "$-a$" if a is a complex number; and (iii) what the *binary*[3] addition operator means—that is, what a+b means for complex numbers a and b. Defining these operators for classes is known as *operator overloading*. We will explain how this is done in C++ below.

---

[2] A unary operator has one input, hence -a is the unary minus operator applied to a.
[3] A binary operator has two inputs, hence a+b is the binary addition operator applied to a and b.

6.4.2 The Class of Complex Numbers

In light of the discussion above, we will write a class of complex numbers with the following members.

- A double precision floating point variable mRealPart containing the real part of the complex number.
- A double precision floating point variable mImaginaryPart containing the imaginary part of the complex number.
- An overridden default constructor ComplexNumber() that initialises the real part and the imaginary part to zero.
- A constructor ComplexNumber(double x, double y) that initialises the real part to x and the imaginary part to y.
- A method CalculateModulus() that returns a double precision floating point variable containing the modulus (or magnitude) of the complex number.
- A method CalculateArgument() that returns a double precision floating point variable containing the argument (or phase) of the complex number.
- A method CalculatePower(double n) that returns the complex number calculated when raising the original complex number to the power n.
- Overloading of the assignment operator.
- Overloading of the unary subtraction operator.
- Overloading of the binary addition and subtraction operators.
- Overloading of the output stream (<<) insertion operator which gives control of the output format for complex numbers.

A suitable header file for this class is shown below. This should be saved as ComplexNumber.hpp. We have made the data associated with each complex number—i.e. the real part and the imaginary part—private members of this class to prevent inadvertent corruption of these members. These members can, of course, be accessed by the methods of the class.

Listing 6.9 The file ComplexNumber.hpp

```
1   #ifndef COMPLEXNUMBERHEADERDEF
2   #define COMPLEXNUMBERHEADERDEF
3
4   #include <iostream>
5
6   class ComplexNumber
7   {
8   private:
9       double mRealPart;
10      double mImaginaryPart;
11  public:
12      ComplexNumber();
13      ComplexNumber(double x, double y);
14      double CalculateModulus() const;
15      double CalculateArgument() const;
16      ComplexNumber CalculatePower(double n) const;
```

```
17    ComplexNumber& operator=(const ComplexNumber& z);
18    ComplexNumber operator-() const;
19    ComplexNumber operator+(const ComplexNumber& z) const;
20    ComplexNumber operator-(const ComplexNumber& z) const;
21    friend std::ostream& operator<<(std::ostream& output,
22                                    const ComplexNumber& z);
23  };
24
25  #endif
```

Code for the class members that are methods is shown in Listing 6.10, and should be saved as `ComplexNumber.cpp`.

In the code in Listing 6.10 we have written two constructors. The first of these (lines 6–10) overrides the automatically generated default constructor, and initialises both the real part and the imaginary part of the complex number to zero if no values are specified. The second constructor (lines 13–17) accepts two double precision floating point variables, sets the real part of the complex number to the first of these, and the imaginary part of the complex number to the second of these. Readers who have followed the discussion of constructors for the class of books will need no more discussion on the implementation of these constructors. We have not defined a new copy constructor because the automatically generated copy constructor behaves correctly.

We now turn our attention to the third method in the code below, the method for calculating the modulus of a complex number in lines 21–25. As this method returns the modulus of the complex number, which is a double precision floating point variable, we begin line 21 with the word `double` to reflect this. This is then followed by the text `ComplexNumber::CalculateModulus()` to indicate that: (i) it is a member of the class `ComplexNumber`; and (ii) the method is called `CalculateModulus`. The text `()` indicates that no arguments are required. Recall that member methods can access all class members, and so there is no need to specify either the real part or the imaginary part of the complex number in the list of arguments. Line 21 then concludes with the reserved keyword `const` to ensure that both the real part and the imaginary part of the complex number whose modulus is being calculated are left unchanged by this method. A simple calculation is then performed to return the modulus of this number. The fourth method in the code above, lines 29–32, uses very similar ideas to calculate the argument of a complex number. Readers should work through this method to ensure that they understand exactly why the function has been written in this way.

Listing 6.10 The file `ComplexNumber.cpp`

```
1  #include "ComplexNumber.hpp"
2  #include <cmath>
3
4  // Override default constructor
5  // Set real and imaginary parts to zero
```

```
6   ComplexNumber::ComplexNumber()
7   {
8      mRealPart = 0.0;
9      mImaginaryPart = 0.0;
10  }
11
12  // Constructor that sets complex number z=x+iy
13  ComplexNumber::ComplexNumber(double x, double y)
14  {
15     mRealPart = x;
16     mImaginaryPart = y;
17  }
18
19  // Method for computing the modulus of a
20  // complex number
21  double ComplexNumber::CalculateModulus() const
22  {
23     return sqrt(mRealPart*mRealPart+
24                 mImaginaryPart*mImaginaryPart);
25  }
26
27  // Method for computing the argument of a
28  // complex number
29  double ComplexNumber::CalculateArgument() const
30  {
31     return atan2(mImaginaryPart, mRealPart);
32  }
33
34  // Method for raising complex number to the power n
35  // using De Moivre's theorem - first complex
36  // number must be converted to polar form
37  ComplexNumber ComplexNumber::CalculatePower(double n) const
38  {
39     double modulus = CalculateModulus();
40     double argument = CalculateArgument();
41     double mod_of_result = pow(modulus, n);
42     double arg_of_result = argument*n;
43     double real_part = mod_of_result*cos(arg_of_result);
44     double imag_part = mod_of_result*sin(arg_of_result);
45     ComplexNumber z(real_part, imag_part);
46     return z;
47  }
48
49  // Overloading the = (assignment) operator
50  ComplexNumber& ComplexNumber::
51                 operator=(const ComplexNumber& z)
52  {
53     mRealPart = z.mRealPart;
54     mImaginaryPart = z.mImaginaryPart;
55     return *this;
56  }
```

```cpp
57
58   // Overloading the unary - operator
59   ComplexNumber ComplexNumber::operator-() const
60   {
61      ComplexNumber w;
62      w.mRealPart = -mRealPart;
63      w.mImaginaryPart = -mImaginaryPart;
64      return w;
65   }
66
67   // Overloading the binary + operator
68   ComplexNumber ComplexNumber::
69                  operator+(const ComplexNumber& z) const
70   {
71      ComplexNumber w;
72      w.mRealPart = mRealPart + z.mRealPart;
73      w.mImaginaryPart = mImaginaryPart + z.mImaginaryPart;
74      return w;
75   }
76
77   // Overloading the binary - operator
78   ComplexNumber ComplexNumber::
79                  operator-(const ComplexNumber& z) const
80   {
81      ComplexNumber w;
82      w.mRealPart = mRealPart - z.mRealPart;
83      w.mImaginaryPart = mImaginaryPart - z.mImaginaryPart;
84      return w;
85   }
86
87   // Overloading the insertion << operator
88   std::ostream& operator<<(std::ostream& output,
89                            const ComplexNumber& z)
90   {
91      // Format as "(a + bi)" or as "(a - bi)"
92      output << "(" << z.mRealPart << " ";
93      if (z.mImaginaryPart >= 0.0)
94      {
95         output << "+ " << z.mImaginaryPart << "i)";
96      }
97      else
98      {
99         // z.mImaginaryPart < 0.0
100        // Replace + with minus sign
101        output << "- " << -z.mImaginaryPart << "i)";
102     }
103     return output;
104  }
```

Much of the discussion on the methods CalculateModulus and Calcu-lateArgument applies to the fifth method in lines 37–47 of the code, namely the function CalculatePower, which is used to return the nth power of a given complex number. We perform this calculation by first writing the complex number in polar form, that is, $z = re^{i}\theta$. We may then write $z^n = r^n e^{in\theta}$, which has real part $r^n \cos(n\theta)$, and imaginary part $r^n \sin(n\theta)$. This method requires some different features to the methods of this class already described, which we now explain. In line 37, we specify that the type of variable returned is of type ComplexNumber: that is, methods can be used to return an instance of a class as well as simpler variable types such as double. This method also requires input of the exponent to which we raise the complex number: this is specified by the "double n" in brackets at the end of line 37. Inside the method, the first two lines of code calculate the modulus and argument of the original number using the two class members CalculateModulus and CalculateArgument—this demonstrates how to call these methods from within the class. The next two lines then perform the calculations required on both the modulus and argument of the complex number to raise it to the power of n. Having set both the real part and the imaginary part of the resulting complex number, this complex number is then returned.

In lines 50–56, we overload the assignment operator. Note that the argument to the assignment operator is a reference to another instance of the class, rather than the object itself. This is because, by default, all method arguments are called by copy, necessitating the overhead of the use of the copy constructor in making the assignment. The use of the const keyword guarantees that the assignment operator will not alter the contents of the object argument z. The remainder of the method for assignment uses an entity called this which does not appear to have been declared. For the purpose of this book, the reader need only know that this is a pointer to the complex number that is returned: it is the contents of this which is returned.

The unary subtraction operator is overloaded in lines 59–65. Line 59 explains that: (i) the return type is a ComplexNumber; (ii) the method is a member of the class ComplexNumber; (iii) defines the operator "-"; (iv) the function requires no input arguments (as specified by the empty brackets); and (v) the original complex number is left unchanged (through use of const). An instance of the class ComplexNumber, called w, is then declared in line 61, and the real part and imaginary part of w are set to the negative of those of the original complex number in lines 62 and 63. Finally, the complex number w is returned.

The binary addition operator is defined in lines 68–75. We begin as usual in lines 68–69 by specifying the return type, the class that the function is a member of, the operator and the input argument. There is only one input argument which is that to the right of the + operator—the class itself is the left operand. We declare an instance of a complex number (line 71), perform the required addition (lines 72–73), and then return the result of this addition (line 74). A similar function overloads the binary subtraction operator in lines 78–85.

The final operator is defined in lines 88–103. This is the output stream (<<) insertion operator. The syntax here is different: the operator is not a member method of the class, but is an external function. This operator uses the friend keyword introduced in Sect. 6.3. By using the `friend` keyword for the operator `<<` in line 21 of the header file for complex numbers, we are telling the computer that, although this operator is not a class member, this operator may access all class members— including private members. When this operator is defined in lines 88–103 of the listing above, we see that we do not make it a class member through `ComplexNumber::`. The function defining this operator takes an output stream (such as `std::cout` or an output stream to a file) and inserts characters into it using the complex number z.

We now demonstrate use of the class of complex numbers in the following code. Recall from earlier that when member methods are called that require no arguments we still need to acknowledge that they are functions by using empty brackets, for example `z1.CalculateModulus()` in line 9 of the code below. Note that we can declare an array of complex numbers: this is shown in line 25 of the listing below where we create an array of complex numbers with two entries. In lines 26–27, we set the first element of this array to the complex number `z1`, and the second element of this array to the complex number `z2`. In lines 28 and 29, we show how to access a friend function of an entry of an array, through printing the complex number that is the second entry of the array of complex numbers to screen.

The files `ComplexNumber.hpp` and `ComplexNumber.cpp` given in Listings 6.9 and 6.10 may be downloaded from http://www.springer.com/book/9783319731315.

```cpp
1  #include "ComplexNumber.hpp"
2
3  int main(int argc, char* argv[])
4  {
5      ComplexNumber z1(4.0, 3.0);
6
7      std::cout << "z1 = " << z1 << "\n";
8      std::cout << "Modulus z1 = "
9                << z1.CalculateModulus() << "\n";
10     std::cout << "Argument z1 = "
11               << z1.CalculateArgument() << "\n";
12
13     ComplexNumber z2;
14     z2 = z1.CalculatePower(3);
15     std::cout << "z2 = z1*z1*z1 = " << z2 << "\n";
16
17     ComplexNumber z3;
18     z3 = -z2;
19     std::cout << "z3 = -z2 = " << z3 << "\n";
20
21     ComplexNumber z4;
22     z4 = z1 + z2;
23     std::cout << "z1 + z2  = " << z4 << "\n";
24
25     ComplexNumber zs[2];
26     zs[0] = z1;
27     zs[1] = z2;
```

```
28      std::cout << "Second element of zs = "
29              << zs[1] << "\n";
30
31      return 0;
32  }
```

6.5 Some Additional Remarks on Operator Overloading

In Sect. 6.4.1, we introduced the concept of operator overloading. This concept was demonstrated in Sect. 6.4.2 using the example class of complex numbers. In this example class, we demonstrated how to overload the assignment operator, and both unary and binary addition and subtraction operators. Many more operators may be overloaded, as will be demonstrated in later chapters. In Sect. 8.1, we show how the square bracket operator may be overloaded. In Sect. 8.3.2, we show how the "less than" operator can be overloaded: extending this to the "greater than" operator, the "less than or equals to" operator, the "greater than or equals to" operator, the "not equal to" operator, and the equality operator then follows the same pattern. In Sect. 10.3.4, we demonstrate how to overload the round bracket operator.

6.6 Tips: Coding to a Standard

Many programming organisations and projects use coding standards in an attempt to ensure that the software written is of an appropriate quality. A famous C++ coding style called JSF (Joint Strike Fighter) was drafted for an international aviation project and has now been adopted by many commercial software houses. Some organisations use automatic checks to ensure that their code complies to the standard (to the extent that employees are reprimanded if their work falls short), while other organisations use the standard as a guideline.

Coding standards are basic rules for programming. Some rules dictate how programs should be laid out (in terms of where comments, new lines and spaces should appear). Other rules are about the naming of variables, classes, functions and methods. Still other rules outlaw various programming practises which, although legal in the language, are considered dangerous (such as returning a pointer to locally allocated memory). The reasons for adopting coding standards are various, but it is generally believed that they promote code which is more reliable, portable, maintainable, readable and extensible.

We believe that a few simple coding rules make programs much more readable (and therefore more maintainable). For this reason, we have used a small set of coding standard rules throughout this book. We don't always follow these rules rigidly, especially when we present small fragments of programs, but once you are familiar with some of the rules we are using then our presentation of code should make more sense.

1. Code within blocks (such as those introduced in Sect. 5.1, as well as functions, loops, branches of if statements, and other places which may have curly brackets) is indented. The curly brackets ({ and }) are always used, even in single-statement blocks (see Sect. 2.1.1), and they appear on a line of their own.
2. Lines of code which are too long to fit comfortably within the width of an editor are split across multiple lines with a suitable indentation.
3. Names for variables and functions are meaningful (e.g., local_index or numberOfNodes) but are not so verbose that they become too long and unwieldy.
4. Variables are declared close to where they are used, rather than at the beginning of a function. This is so that the context is clear (see Sect. 5.1). Loop counter variables are declared in the context of the loop, that is, we write

```
1   for (int i=0; i<10; i++)
2   {
3       std::cout << i << "\n";
4   }
```

rather than

```
1   int i;
2   for (i=0; i<10; i++)
3   {
4       std::cout << i << "\n";
5   }
```

5. Locally declared variable names have underscores (e.g., total_sum).
6. Where types are pointers or references the "*" or "&" character is written adjacent to the native type, with no space between, that is,

```
    int* i;
```

rather than

```
    int *i;
```

As explained in Sect. 4.1.2, a consequence of this rule is that each pointer variable declaration should appear on its own line.

7. Pointer names begin with "p" (e.g., `p_return_result` or `pLastResult`). One exception to this rule is when the pointer is used for an array of values stored in dynamically allocated memory.

8. Function names are in camel-case (i.e., where capital letters begin each word) and the first word is a verb, to indicate what it is that they *do* (e.g. `GetSize()` or `InitialisePreconditioner()`). This applies to class methods as well as to regular functions.

9. Names of arguments to functions (and class methods) are in also camel-case, but they begin in lower-case (e.g., `firstDimension`). The same format is also applied to member data of classes, but the following rule helps us to distinguish them.

10. Class data which have access controls are also in camel-case with "m" (for "my") to denote "private" or "protected" (e.g., `mSize` or `mpQuadraticMesh` where the latter is a private pointer). Since it is advisable for member data to be private, this naming convention allows us to distinguish, in the body of a class method, between the method arguments and the class variables.

11. Class names are also in camel-case (as are function names), but they can be distinguished by the context (e.g., `FiniteElementSolver` or `PopSinger`).

12. There should be lots of descriptive comments as discussed in Sect. 5.10.

6.7 Exercises

In all of the exercises below, test your code using suitably chosen test cases.

6.1 The files `ComplexNumber.hpp` and `ComplexNumber.cpp` given in Listings 6.9 and 6.10 may be downloaded from http://www.springer.com/book/9783319731315. Extend this class to include the following features.

1. Methods called `GetRealPart` and `GetImaginaryPart` that allow us to access the corresponding private members. In the class of complex numbers, the members representing the real and imaginary parts of the complex number— called `mRealPart` and `mImaginaryPart`—are private members. These members may be set through using a constructor, but there is no way to access them.

2. Friend functions `RealPart` and `ImaginaryPart` so one may either write `z.GetImaginaryPart()` or `ImaginaryPart(z)`.

3. An overridden copy constructor.

4. A constructor that allows us to specify a real number in complex form through a constructor that accepts one double precision floating point variable as input, sets the real part of the complex number to the input variable, and the imaginary part to zero.

5. A const method `CalculateConjugate` which returns the complex conjugate $x - iy$ of a complex number $x + iy$.

6. A method `SetToConjugate` which has a void return type and sets the complex number $x + iy$ to its complex conjugate $x - iy$.
7. Write code to dynamically allocate memory for a 3×3 matrix of complex numbers. Extend this code to calculate the exponential of the matrix, where the exponential of a matrix A is given by

$$\exp(A) = \sum_{n=0}^{\infty} \frac{A^n}{n!},$$

where, in practice, the infinite sum above is truncated at a suitably large value of n. Having allocated the memory for this array dynamically what should you now do? See Sect. 4.3.3 if you don't know.
8. Test the class to ensure that special cases give sensible results. For example $(0+0i)^n$ should equal zero for most values of n, but any number raised by $n = 0$ should return 1.

6.2 Develop a class of 2×2 matrices of double precision floating point variables that has the features listed below.
1. An overridden default constructor that initialises all entries of the matrix to zero.
2. An overridden copy constructor.
3. A constructor that specifies the four entries of the matrix and allocates these entries appropriately.
4. A method (function) that returns the determinant of the matrix.
5. A method that returns the inverse of the matrix, if it exists.
6. Overloading of the assignment operator, allowing us to write code such as A = B; for instances of the class A and B.
7. Overloading of the unary subtraction operator, allowing us to write code such as A = -B; for instances of the class A and B.
8. Overloading of the binary addition and subtraction operators, allowing us to write code such as A = B + C; or A = B - C; for instances of the class A, B and C.
9. A method that multiplies a matrix by a specified double precision floating point variable.

Inheritance and Derived Classes

In Sect. 6.1.1, we explained how object-oriented programming allowed for a more reliable programming paradigm than was possible using modules. One reason for this, which we touched on briefly, is the availability of *inheritance*. Inheritance allows us to extend the functionality of a class by introducing a new class, known as the *derived class*, that contains all the features of the original class, known as the *base class*.

7.1 Inheritance, Extensibility and Polymorphism

Perhaps the most important feature of object-oriented programming is *inheritance*. This concept allows the functionality of classes to be built into a "family tree". The data, operation and functionality of a given class (the base class, sometimes called the parent class) may be directly reused, extended and modified in another class (the derived or child class). The operation of one base class can be inherited by several derived classes.[1] In turn, these derived classes may become the base classes of further inheritance, giving rise to further generations.

Suppose we have written a class that allows us to solve linear systems. Suppose further that we now want to write a class for solving linear systems that may be used only when the matrix in the linear system is symmetric and positive definite, thus

---

[1]A feature of C++ is that it also allows *multiple inheritance*, not available in other object-oriented languages, where derived classes may inherit from more than one base class. This feature causes some seasoned C++ programmers difficulty, and hence is beyond the scope of this book, although we do briefly discuss this topic in Appendix B.

© Springer International Publishing AG, part of Springer Nature 2017   129
J. Pitt-Francis and J. Whiteley, *Guide to Scientific Computing in C++*, Undergraduate Topics in Computer Science,
https://doi.org/10.1007/978-3-319-73132-2_7

allowing us to solve the system using the very effective conjugate gradient technique discussed in Sect. A.2.3. Much of the functionality required—such as specifying the vectors, matrix and tolerance, and providing a function for calculating the scalar product between two vectors—will already be implemented in the class that has been written to solve more general linear systems. Inheritance allows us to write a new class for solving a special category of linear systems that uses—or inherits—all features of the class for solving general linear systems. If we wanted to extend the functionality of the class that uses the conjugate gradient scheme to include Successive Over–Relaxation (SOR),[2] we simply inherit again so that the SOR variant is a grandchild derived class of the original.

Inheritance gives rise to two important concepts first mentioned in Sect. 1.1.1: *extensibility* and *polymorphism*. Extensibility is the idea, not just that the code can be extended, but that it can be extended easily, and without changing any of the original functional behaviour of the base class. Polymorphism is the ability to perform the same operations on a wide variety of different types of objects. So, for example, the Solve method of the generic linear solver outlined above will perform a certain set of operations. This method of the base class is then redefined in a derived class for symmetric, positive definite matrices, without changing its arguments. At run-time, the program is able to detect which object it has and therefore which version of Solve to run. This version of polymorphism is also known as dynamic polymorphism or run-time polymorphism.

7.2   Example: A Class of E-books Derived from a Class of Books

We now demonstrate the basic features of inheritance through extending the class of books developed in Sect. 6.2. Suppose the owner of a bookshop also runs a website where she not only sells traditional (paper) books, but also electronic e-books. The advantage of the e-book over a traditional book is that it does not need to be parcelled up and sent through the mail. The e-book may be delivered by giving the customer access to a private URL from which they may download it. The bookseller may wish to update her computer system so that a URL attribute is added to each instance of her e-books. She could do this by deriving a class Ebook from the class Book given in Listings 6.6 and 6.7. The class Ebook will have the same members as the class Book, but with two differences. The first difference is that the class member format will be set to "electronic". The second difference is that instances of the class Ebook will have an additional class member hiddenUrl that contains the private URL. The header file for this class is given below.

---

[2]SOR is an iterative technique for solving linear systems: see, for example, Iserles [1].

As the class `Ebook` is derived from the class `Book`, we include the header file for the class `Book` in the header file for the class `Ebook` below. Line 7 of this listing specifies that the class `Ebook` is indeed derived from the class `Book`, and the word "`public`" in this line has the effect that:

1. public members of `Book` are public members of `Ebook`;
2. protected members of `Book` are protected members of `Ebook`; and
3. private members of `Book` are hidden from `Ebook`, and so may not be used by the derived class.

This is known as *public inheritance*. We will discuss access privileges for derived classes in more detail in Sect. 7.3.

Listing 7.1 The file `Ebook.hpp`

```
1   #ifndef EBOOKHEADERDEF
2   #define EBOOKHEADERDEF
3
4   #include <string>
5   #include "Book.hpp"
6
7   class Ebook: public Book
8   {
9   public:
10     Ebook();
11     std::string hiddenUrl;
12  };
13
14  #endif
```

Based on the discussion above, all public and protected members of the class `Book` defined in Listing 6.6 are available to instances of the class `Ebook`. This has the possibly unintended effect that the member `mYearOfPublication` is not directly available to the derived class `Ebook`, as this member is private and therefore not available to the derived class. This member is, however, still available indirectly through the public methods of the base class `SetYearOfPublication` and `GetYearOfPublication`—as these members are public they are available to the derived class, and can be used to access the member `mYearOfPublication`. The other difference between the derived class and the parent class is that we have declared two additional members in the listing above: an overridden default constructor, and a string member representing the hidden URL.

The overridden default constructor is given below, where the format is set to "electronic" as required. Note the syntax for overridden default constructors below: this allows the default constructor for the base class `Book` to be called first, setting the author, the title, and the publisher to "unspecified". The format is then set to "electronic" inside the overridden default constructor for the derived class.

```
1  #include "Ebook.hpp"
2
3  Ebook::Ebook() : Book()
4  {
5    format = "electronic";
6  }
```

Example code using the class Ebook is given below. Note that the member format of an instance of the class Ebook is automatically set to electronic.

```
1  #include <iostream>
2  #include "Ebook.hpp"
3
4  int main(int argc, char* argv[])
5  {
6    Ebook holiday_reading;
7    holiday_reading.title = "The skull beneath the skin";
8    holiday_reading.author = "P D James";
9    std::cout << "The author is " << holiday_reading.author
10             << "\n";
11   std::cout << "The title is " << holiday_reading.title
12             << "\n";
13   std::cout << "The format is " << holiday_reading.format
14             << "\n";
15
16   holiday_reading.SetYearOfPublication(1982);
17   std::cout << "Year of publication is "
18             << holiday_reading.GetYearOfPublication()
19             << "\n";
20
21   holiday_reading.hiddenUrl =
22             "http://ebook.example.com/example-book";
23   std::cout << "The URL is "
24             << holiday_reading.hiddenUrl << "\n";
25
26   return 0;
27 }
```

Figure 7.1 shows, in schematic form, a representation of how the class Ebook relates to its parent class Book. This representation is given in the *Unified Modelling Language* (*UML*) format where each class is shown as a box. Space inside each box is divided into three components: the class name, a list of the data contained in the class and a list of the class methods. A + sign signifies data and methods which are public. Private data or methods (mYearOfPublication in this case) carry a − sign, while protected members would be given a # sign.

Fig. 7.1 An inheritance
graph, showing that Ebook is
derived from the Book base
class

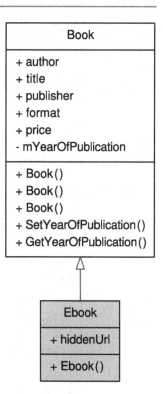

The arrow between the boxes shows the child–parent inheritance relationship. The reason for the repetition of "+ Book()" in the base class is to show that Book has three different constructors: the default constructor, the copy constructor and a specialised Book constructor for setting the title attribute. These three constructors were introduced in Sect. 6.2.7. Ebook has only one constructor which is the overridden default (no argument) constructor given above which sets the format attribute.

7.3 Access Privileges for Derived Classes

When developing a class, we specify all class members as being public, protected or private members. When a class is derived from this base class, we need to know what access privileges the members of the base class have in the derived class. In the class Ebook that we derived from the class Book in Sect. 7.2, we used public inheritance in line 7 of Listing 7.1. There are two other types of inheritance: *protected inheritance*; and *private inheritance*. These three different types of inheritance determine the access privileges of the base class members in the derived class. In Table 7.1, we state these access privileges.

Table 7.1 Access privileges for derived classes

Access privilege in base class	Type of inheritance		
	Public	Protected	Private
Public	Public	Protected	Private
Protected	Protected	Protected	Private
Private	Hidden	Hidden	Hidden

7.4   Classes Derived from Derived Classes

We may derive classes from classes that are themselves derived classes, as discussed in Sect. 7.1. If Class2 is derived from Class1, we may derive a new class Class3 from Class2 in exactly the same way as in Sect. 7.2, as shown in the header file for Class3 shown below.

```
1   #ifndef CLASS3DEF
2   #define CLASS3DEF
3
4   #include "Class2.hpp"
5
6   class Class3: public Class2
7   {
8   public:
9     double newMember;
10  };
11
12  #endif
```

7.5   Run-Time Polymorphism

Polymorphism may be used when a number of classes are derived from the base class, and for some of these derived classes we want to override one—or more—of the methods of the base class. Suppose we have developed a class of guests who stay at a hotel. This class will include members such as name, room type, arrival date, number of nights booked, and a member method that computes the total bill. It is likely that the hotel has negotiated special nightly rates for individuals from particular organisations. To reflect this, the method that computes the total bill must act differently on guests from these organisations. This may be incorporated into software in a very elegant manner through the use of *virtual methods* where the method does different things for different derived classes. This is implemented by the use of the virtual keyword, shown in the header file for the class of hotel guests shown below. The virtual keyword is a signal to the compiler that a method has the potential to be overridden by a derived class.

```
1   #ifndef GUESTDEF
2   #define GUESTDEF
3
4   #include <string>
5
6   class Guest
7   {
8   public:
9     std::string name, roomType, arrivalDate;
10    int numberOfNights;
11    double telephoneBill;
12    virtual double CalculateBill();
13  };
14
15  #endif
```

The implementation of the method CalculateBill is given in the listing
below, where the total bill is given by multiplying the number of nights that a guest
stayed in the hotel by a nightly rate of £50, and adding the telephone bill to this
figure. Even though this method is a virtual method, it is written in exactly the same
way as if it were not declared as virtual.

```
1   #include "Guest.hpp"
2
3   double Guest::CalculateBill()
4   {
5     return telephoneBill + ((double)(numberOfNights))*50.0;
6   }
```

Suppose now that the hotel have negotiated a deal with a company that reduces
the room rate to £45 for the first night that a guest stays, and £40 for subsequent
nights, and offers free telephone calls. This may be implemented by deriving a class
SpecialGuest from the class Guest as shown below.

```
1   #ifndef SPECIALGUESTDEF
2   #define SPECIALGUESTDEF
3
4   #include "Guest.hpp"
5
6   class SpecialGuest: public Guest
7   {
8   public:
9     double CalculateBill();
10  };
11
12  #endif
```

The method `CalculateBill` for this derived class is then implemented using the code below.

```
#include "SpecialGuest.hpp"

double SpecialGuest::CalculateBill()
{
  return 45.0 + ((double)(numberOfNights-1))*40.0;
}
```

Note that declaring the member method `CalculateBill` as virtual in the class `Guest` does not require that the method must be overridden (redefined) in derived classes: it simply gives us the option to override it.

The real power of run-time polymorphism can be seen when we use only pointers to the *base class* in a family tree of objects. It might not be obvious what the exact type of each object in our program is, but the run-time system is able to find out. In the following code, there are three pointers to `Guest` objects, but one of them is in actuality a `SpecialGuest` and therefore has a reduced bill. One might imagine a larger-scale program running over an array of `Guest` pointers—representing those guests who are checking out—each of which has their own mechanism for calculating the bill. The programmer does not need to be aware which of these `Guest` objects might be actually be a `SpecialGuest`.[3]

```
#include <iostream>
#include "Guest.hpp"
#include "SpecialGuest.hpp"

int main(int argc, char* argv[])
{
  Guest* p_gu1 = new Guest;
  Guest* p_gu2 = new Guest;
  Guest* p_gu3 = new SpecialGuest;

  //Set the three guests identically
  p_gu1->numberOfNights = 3;
  p_gu1->telephoneBill = 0.00;
  p_gu2->numberOfNights = 3;
  p_gu2->telephoneBill = 0.00;
  p_gu3->numberOfNights = 3;
  p_gu3->telephoneBill = 0.00;

  std::cout << "Bill for Guest 1 = "
            << p_gu1->CalculateBill() << "\n";
```

[3]The advanced programmer can test if a `Guest` is a `SpecialGuest` using a feature called *dynamic casting*.

```
21    std::cout << "Bill for Guest 2 = "
22              << p_gu2->CalculateBill() << "\n";
23    std::cout << "Smaller bill for Guest 3 = "
24              << p_gu3->CalculateBill() << "\n";
25    delete p_gu1;
26    delete p_gu2;
27    delete p_gu3;
28    return 0;
29 }
```

7.6 The Abstract Class Pattern

Suppose we want to write an object-oriented program for calculating the numerical solution of initial value ordinary differential equations of the form

$$\frac{dy}{dt} = f(t, y), \qquad y(T_0) = Y_0,$$

where $f(t,y)$ is a given function, and T_0, Y_0 are given values. Many methods exist for calculating the numerical solution of equations such as these, for example, the forward Euler method, Heun's method, various Runge–Kutta methods, and various multistep methods. One way of implementing these numerical methods would be to write a class called AbstractOdeSolver that has members that would be used by all of these numerical methods, such as variables representing the stepsize and initial conditions, a method that represents the function $f(t, y)$ on the right-hand side of the equation above, and a virtual method SolveEquation for implementing one of the numerical techniques described above. We would then implement each of the numerical methods using a class derived from AbstractOdeSolver, and overriding the virtual function SolveEquation. The derived classes would then contain members that allow a specific numerical algorithm to be implemented, as well as the members of the base class AbstractOdeSolver that would be required by all of the numerical solvers.

Using the class structure described above, the base class AbstractOdeSolver would not actually include a numerical method for calculating a numerical solution of a differential equation, and so we would not want to ever create an instance of this class. We can automatically enforce this by making AbstractOdeSolver an *abstract class*. This is implemented by setting the virtual functions Solve-Equation and RightHandSide to be *pure virtual functions* as shown in lines 15 and 16 of the listing for AbstractOdeSolver.hpp below. We indicate that these functions are pure virtual functions by completing the declaration of these members with "= 0" as shown in the listing below. Should we mistakenly attempt to create an instance of the class AbstractOdeSolver we would get a compilation error. An investigation into pure virtual functions is made in Exercise 7.2.

Listing 7.2 The file `AbstractOdeSolver.hpp`

```
 1  #ifndef ABSTRACTODESOLVERDEF
 2  #define ABSTRACTODESOLVERDEF
 3
 4  class AbstractOdeSolver
 5  {
 6  protected:
 7    double stepSize;
 8    double initialTime;
 9    double finalTime;
10    double initialValue;
11  public:
12    void SetStepSize(double h);
13    void SetTimeInterval(double t0, double t1);
14    void SetInitialValue(double y0);
15    virtual double RightHandSide(double y, double t) = 0;
16    virtual double SolveEquation() = 0;
17  };
18
19  #endif
```

A class is an abstract class if it contains one or more pure virtual methods. We do not discuss implementation of the class `AbstractOdeSolver` or the derived classes further here: these classes are developed in the exercises at the end of this chapter.

7.7   Tips: Using a Debugger

In Sect. 1.7 we gave a few tips about how to debug your code using simple techniques such as printing information out to the screen, and we also promised to give a little more information on using a debugger to inspect your code. There is a wide-range of open source and commercial tools to support you, should you wish to do this.

The easiest debuggers to use are those which are integrated with your development environment (such as Visual Studio or Eclipse). These integrated debuggers allow you to set breakpoints (places where you wish to temporarily pause execution) by clicking and selecting individual lines of code in your editing window. In the case of Eclipse, the debugging options basically provide a point and click front-end interface on top of a less user-friendly text-based debugger such as `gdb`.

The next level of sophistication is a graphical standalone debugger. Many of those available are actually a front-end to a text-based debugger, whereas some, such as `ups` are completely self-contained debuggers. A popular open source graphical front-end debugger is `ddd` which is a graphical interface to `gdb`, although it can also interface with a range of low-level debugging tools for a variety of programming languages. There are many other graphical front-end debuggers available such as `KDbg` and `Xxgdb`.

The lowest level of sophistication is the text-based debugger. The most widely used of these is the open source GNU debugger gdb, but many commercial compilers offer their own debugging environments.

All the debugging tools mentioned will allow you to walk through the code line by line, function call by function call, or to the next break point. If your program aborts with a *segmentation fault*, then the debugger will stop at the place where the fault happened, allowing you to see the line which caused the error. At any stage in execution, you will be able to inspect the values of the program variables and classes. You will also be able to inspect the *back-trace* (or *stack*) which shows the function calling sequence which led from the main function to a particular line of code.

Our advice is to debug your code with a graphical front-end to gdb, such as the popular ddd. Such tools are easy to download and install. The fact that they have a graphical interface with a built-in help system will allow you to rapidly see what the capabilities are. We also need to stress at this point that debuggers do not cope well with optimised code. Before you load the program into the debugger, you must remember to first compile your code with the "-g" flag (see Sect. 1.3.3).

7.8 Exercises

7.1 In this question, we will develop classes to describe the students at a university.
1. Write a class of students at the university that has the following public members:
   - a string for the student's name;
   - a double precision floating point variable that stores the library fines owed by the student;
   - a double precision floating point variable that stores the tuition fees owed by the student;
   - a method that returns the total money owed by the student, that is, the sum of the library fines and tuition fees associated with a given student;
   - a few constructors that take different arguments.
2. The library fines owed by the students must be a nonnegative number. Enforce this by making a student's library fines a private member of the class. Write one method that allows the user to set this variable only to nonnegative values, and another method that can be used to access this private variable. Both methods should be public members of the class.
3. Students at the university are either graduate students or undergraduate students. All undergraduate students are full-time students. Graduate students may be full-time students or part-time students. Derive a class of graduate students from the class of students that you have already written with an additional member variable that stores whether the student is full-time or part-time.
4. Graduate students do not pay tuition fees. Use polymorphism to write a method that calculates the total money owed by a graduate student. This will require the method for calculating the total money owed to be a virtual function of the parent class.

5. Ph.D. students are a special class of graduate students who do not pay library fines. Derive a class of Ph.D. students from the class of graduate students. Write a method that calculates the total money owed by a Ph.D. student.

7.2 This exercise is an investigation into proper use of the virtual keyword and into safe ways of making *abstract* classes.

The following program presents a small hierarchy of classes using the *abstract class pattern* described in Sect. 7.6. There is an abstract class AbstractPerson, which is intended never to be instantiated, and two derived classes, Mother and Daughter. The code in the main function demonstrates the power of polymorphic inheritance. It shows that it is possible to have a variety of objects of the same family stored as pointers to a generic abstract type, each of which could be a different concrete class. The AbstractPerson class promises a Print method, but it is only at run-time that the system inspects the class pointed to by p_mother and works out which Print method to invoke.

```
 1  #include <iostream>
 2
 3  class AbstractPerson
 4  {
 5  public:
 6    virtual void Print(){std::cerr<<"Never instantiate\n";}
 7  };
 8
 9  class Mother : public AbstractPerson
10  {
11  public:
12    virtual void Print(){std::cout<<"Mother\n";}
13  };
14
15  class Daughter : public Mother
16  {
17  public:
18    void Print(){std::cout<<"Daughter\n";}
19  };
20
21  int main(int argc, char* argv[])
22  {
23    AbstractPerson* p_mother = new Mother;
24    AbstractPerson* p_daughter = new Daughter;
25    p_mother->Print();
26    p_daughter->Print();
27    delete p_mother;
28    delete p_daughter;
29  }
```

1. Copy, save, compile and run the above program. The output from the `Print` method calls in lines 25 and 26 ought to be:

```
1  Mother
2  Daughter
```

2. Investigate what happens if you remove the `public` keyword from the inheritance declaration of either derived class (lines 9 and 15). This will make the base class inaccessible from the derived class.

3. Investigate what happens if you remove either of the `virtual` keywords in lines 6 and 12. Also investigate adding the `virtual` keyword on line 18. How does the output change after each of these changes?

4. What happens if you use the code fragment below to instantiate an instance of the abstract class in the main function?

```
29  AbstractPerson* p_abstract = new AbstractPerson;
30  p_abstract->Print();
31  delete p_abstract;
```

5. The *preferred method* of making an abstract class with a pure virtual method (so that it cannot be instantiated) is to give no implementation of that method in the class. This is done by replacing line 6 with the rather strange syntax which was introduced in the `AbstractOdeSolver` of Sect. 7.6:

```
5  public:
6      virtual void Print() = 0;
```

6. After making the `Print` method of `AbstractPerson` pure virtual as above, repeat the exercise in part 3 of removing the virtual keywords in lines 6 and 12.

7. Also after making the method `AbstractPerson::Print()` pure virtual as above, repeat the exercise in part 4 of attempting to instantiate an instance of the abstract class.

7.3 In Sect. 7.6, we discussed how abstract classes could be used to write a library for calculating the numerical solution of initial value ordinary differential equations, i.e. ordinary differential equations of the form

$$\frac{dy}{dt} = f(t, y),$$

for some user specified function $f(t, y)$, where $y = Y_0$ at $t = T_0$ for an initial value Y_0 at some initial time T_0. We want to calculate a numerical solution in the time interval

$T_0 < t < T_1$ where T_1 is the final time. To solve this equation numerically, we require the user to specify an integration step size, which we denote by h. A large variety of numerical methods exist for solving equations such as these and in Sect. 7.6 we explained that, as these methods all required very similar inputs, they could be coded very effectively using an abstract class pattern. We will base the library developed in this exercise on the abstract class in Listing 7.2: you should save this file, and ensure that you understand how the class members relate to the discussion above.

In this exercise, we will develop the library to allow you to solve initial value ordinary differential equations using two methods: the forward Euler method; and a Runge–Kutta method. Using a step size h, we define the points t_i, $i = 0, 1, 2, \ldots, N$ by

$$t_i = T_0 + ih,$$

where h is chosen so that $t_N = T_1$. The numerical solution at these points is denoted by y_i, $i = 0, 1, 2, \ldots, N$. These values of y_i are determined by the numerical technique chosen.

- For the forward Euler method, we set $y_0 = Y_0$. For $i = 1, 2, \ldots, N$, y_i is given by

$$y_i = y_{i-1} + h\, f(t_{i-1}, y_{i-1}).$$

- For the fourth order Runge–Kutta method, we set $y_0 = Y_0$. For $i = 1, 2, \ldots, N$, we calculate y_i using the following formulae:

$$k_1 = hf(t_{i-1}, y_{i-1}),$$
$$k_2 = hf\left(t_{i-1} + \frac{1}{2}h, y_{i-1} + \frac{1}{2}k_1\right),$$
$$k_3 = hf\left(t_{i-1} + \frac{1}{2}h, y_{i-1} + \frac{1}{2}k_2\right),$$
$$k_4 = hf(t_{i-1} + h, y_{i-1} + k_3),$$
$$y_i = y_{i-1} + \frac{1}{6}(k_1 + 2k_2 + 2k_3 + k_4).$$

More details on numerical methods for initial value problems may be found in Kreyszig, [2].

1. Write the methods associated with the class `AbstractOdeSolver` and save these as the file `AbstractOdeSolver.cpp`. Note that you do not have to write the pure virtual functions, as the "= 0" when they are declared in the file `AbstractOdeSolver.hpp` means that these are already written.

2. Derive a class called `FowardEulerSolver` that allows the user to specify the function `RightHandSide`, and contains a method `SolveEquation` that uses the forward Euler method to calculate the values of y_i as described above, and writes the values of t_i and y_i to file. You may want to refer back to Sect. 5.7 to remind yourself how to allow a user to specify a function.

3. Test the class `FowardEulerSolver` using the initial value ordinary equation

$$\frac{dy}{dt} = 1 + t,$$

for the time interval $0 < t < 1$, and with initial condition $y = 2$ at $t = 0$. This equation has solution $y = (t^2 + 2t + 4)/2$. Investigate how the choice of step size affects the accuracy of the solution.

4. Repeat the two sub-parts above using the fourth order Runge–Kutta method to calculate the values of y_i.

Templates

8

If we want to write a function that returns the larger of two numbers, and we want this function to be used for both integer variables and double precision floating point variables, then we could use function overloading and write two functions: one for integer variables and the other for double precision floating point variables. Both of these functions would require only a few lines of code, and it would not be difficult to maintain both functions. For larger functions maintaining more than one function to do the same operations may be problematic. This may be avoided by the use of *templates*, a feature of the C++ language that allows very general code to be written.

We begin this chapter by discussing templates and the flexibility that they permit. One library associated with C++ is the *Standard Template Library* (STL): we conclude this chapter by giving a brief survey of this library, and other functionality that has been introduced in recent C++ standards.

8.1 Templates to Control Dimensions and Verify Sizes

Many scientific computing applications are underpinned by vectors and matrices. We have seen earlier that these are represented in C++ by arrays. Under normal circumstances there is no check, when we attempt to access elements of an array, that the index is a valid index. For example, in the code fragment below we attempt to access the element with index 7 when the array only has 5 elements. Although this is clearly an error, it may not trigger a compiler or run-time error. The most likely outcome when code including these lines is executed is a segmentation fault or an incorrect answer.

© Springer International Publishing AG, part of Springer Nature 2017   145
J. Pitt-Francis and J. Whiteley, *Guide to Scientific Computing*
in C++, Undergraduate Topics in Computer Science,
https://doi.org/10.1007/978-3-319-73132-2_8

```
1    double A[5];
2    A[7] = 5.0;
```

If this fragment is part of a large program, it could be difficult to locate this error. It would be therefore be useful if we could use arrays with an additional feature that a check for validity of the index is performed each time an element of the array is accessed. This may be achieved using the class shown below, which is referred to as a *templated class*.

Listing 8.1 `DoubleVector.hpp`

```
1    #include <cassert>
2
3    template<unsigned int DIM> class DoubleVector
4    {
5    private:
6      double mData[DIM];
7
8    public:
9      double& operator[](int index) // overloading the []
10                                    // operator
11     {
12        assert(index < DIM);
13        assert(index > -1);
14        return(mData[index]);
15     }
16   };
```

The class in the listing above allows us to declare instances of `DoubleVector`, specifying the length of the array. The entries of the array are private members of this class and so can't be accessed in the normal way that we would access elements of an array. Instead we access members of this class by overloading the square bracket operator. Overloading this operator allows us to check that the index is a valid index before returning the variable requested.

Use of the class above is demonstrated in the code below. Note (in line 6) how using this class requires us to declare the array v as an instance of a `DoubleVector`, with the size of this array being enclosed within pointed brackets. Subsequently this array is accessed in exactly the same way as a normal array, but with the additional feature that a check is carried out on the index every time an element of the array is accessed through the overloading of the square bracket operator.

Listing 8.2 `UseDoubleVector.cpp`

```
1   #include <iostream>
2   #include "DoubleVector.hpp"
3
4   int main(int argc, char* argv[])
5   {
6     DoubleVector<5> v;
7     v[0] = 1.0; // This is OK
8     v[7] = 5.0; // Will trip assertion
9
10    return 0;
11  }
```

8.2 Templates for Polymorphism

There are very good reasons in C++, and many other programming languages, for distinguishing between integer variables and floating point variables. For example, the argument(s) used to access an element of an array may only take integer values which provides one level of validation that the index is correct. Furthermore, integers may be stored much more efficiently than floating point variables. One slight drawback in having to distinguish between these variables is that if we want to write a function that is valid for all numerical variables—that is, both integers and floating point variables—we have to write more than one instance of the same function. Templates, however, provide a way around this.

The program below demonstrates how a function `GetMaximum` that returns the maximum of two numbers, either integers or floating point variables, may be written. The code is very similar to the code that we would write to calculate the maximum of two numbers, although there are two important differences. The first difference is that the function prototype in line 3 of the listing specifies that the function is defined for a general class T, and that the return type and both function arguments will be instances of the same class T. To call the function, we have to put the data type used in angled brackets as is shown in lines 7 and 8 of the listing. The function `GetMaximum` demonstrates *polymorphism*, because it can perform the same operation on different types of input argument. This type of polymorphism is also called static polymorphism or compile-time polymorphism, because when the compiler sees line 7 or 8 of the listing it makes a specific version of `GetMaximum` ready for the int or double type.

```
1   #include <iostream>
2
3   template<class T> T GetMaximum(T number1, T number2);
4
5   int main(int argc, char* argv[])
```

```
6    {
7      std::cout << GetMaximum<int>(10, -2) << "\n";
8      std::cout << GetMaximum<double>(-4.6, 3.5) << "\n";
9
10     return 0;
11   }
12
13   template<class T> T GetMaximum(T number1, T number2)
14   {
15     T result;
16     if (number1 > number2)
17     {
18       result = number1;
19     }
20     else
21     {
22       //number1 <= number2
23       result = number2;
24     }
25     return result;
26   }
```

8.3 A Brief Survey of the Standard Template Library

The Standard Template Library (STL) contains many commonly used patterns that may be reused for different types of objects. In this survey, we give a summary of the features available that are particularly relevant to writers of scientific software.

Containers, such as random-access vectors and sets, are dynamic arrays where the STL is responsible for memory management. We now demonstrate how these two containers may be used. Other containers that are available in the STL are maps, multimaps, multisets, lists and deques (double-ended queues, pronounced "decks"). There are also many more algorithms that may be performed on these containers other than those presented here. Some of these other containers and algorithms do not have application in scientific computing software and so we do not discuss them here. Nevertheless, it is useful for readers to be aware that they exist.

8.3.1 Vectors

The STL vector class is a very useful container because it is an *extensible* class which has a similar interface to the regular C++ array. The fact that it is extensible means that its size is not fixed (either at compile time or at the time that it is created) and that it will grow to accommodate new items as necessary. One can either declare an empty STL vector of minimal capacity which then grows by adding new items to it.

or one can exploit efficiency savings by knowing the maximum size at compile time or run time.

If you explore available STL containers, you will notice that the interface for the STL vector is very similar to the interface for the other basic container types *deque* and *list*. This is a good example of object *abstraction*, because the details which distinguish these container types from each other are not exposed to the user. The main differences between these types of containers are in the efficiency which STL guarantees for various operations: it is possible to retrieve an item from an STL vector via its index in a single operation, but this is not possible from an STL list. It is generally only efficient to insert and delete elements to the back of a vector object and to the front or back of a deque. The list type allows efficient constant time insertion and deletion anywhere in the container.

The use of the vector container is shown in the listing below. Several features of the STL are included in this listing which we now highlight.

- To use the vector container, we must include the `vector` header file (line 2). For some algorithms that may be used on STL vectors, such as sorting, we must include the `algorithm` header file (line 3).
- In line 8, we declare a vector of strings called `destinations`. Note that we do not have to state the size of the vector: the STL will handle this for us. We can write `std::vector<std::string> destinations(50);` if we wished to begin with a vector of 50 empty strings rather than an empty vector.
- In line 9, we *reserve* 6 elements. This sets the vector's *capacity* without changing the number of items in the vector. Although this line is unnecessary, it may produce efficiency savings in more memory-intensive code because it establishes that 6 items can be stored in the vector without having to reallocate any memory later.
- In line 10, we introduce our first entry to the vector, the string "Paris". The member function `push_back` appends a copy of this string to the current vector, which is currently empty.
- In line 11, we append another entry to the end of the vector, that is, the second entry of this vector is "New York".
- In line 12, we append a further entry to the vector, that is, the third entry of this vector is "Singapore".
- In lines 13 and 14, we demonstrate the use of the member function `size` for accessing the number of elements of the vector.
- In lines 17–20, we show that entries of the vector may be accessed in the same way as for a standard vector.
- Lines 22–26 demonstrate how to access entries of the vector using an *iterator*. The iterator is declared in line 22, where we define what type of vector the iterator is associated with, that is, in this case a vector of strings. In line 23, we construct a `for` loop that iterates from the start of the vector to the end of the vector using this iterator. The entries are printed using line 25, which prints out the contents of the vector entry that the iterator is pointing at. Note the use of the overloaded `*` operator which looks like a pointer de-reference.
- In line 28, we add a string to the *start* of a vector by using the `insert` method, and inserting at the start of the vector using the `begin` method: all subsequent entries are now moved one place back.

- In line 29, we add a string to the vector, and place it in the second position: all subsequent entries are again moved one place back.
- In line 30, we add another entry to the end of the vector. We then print out the number of entries of the vector, and the entries, using lines 31–38.
- In lines 40 and 41, we erase all entries of the vector that appear after the third entry, and then print out the number of entries of the vector, and the entries, using lines 42–49.
- In line 51, we use the algorithm `sort`: this algorithm will sort a vector of strings into alphabetical order and requires the header file `algorithm` as described above. This is verified by printing the entries of the vector using lines 52–59.

Listing 8.3 Example use of `std::vector`

```
1   #include <iostream>
2   #include <vector>
3   #include <algorithm>
4   #include <string>
5
6   int main(int argc, char* argv[])
7   {
8     std::vector<std::string> destinations;
9     destinations.reserve(6);
10    destinations.push_back("Paris");
11    destinations.push_back("New York");
12    destinations.push_back("Singapore");
13    std::cout << "Length of vector is "
14            << destinations.size() << "\n";
15    std::cout << "Entries of vector are\n";
16
17    for (int i=0; i<3; i++)
18    {
19      std::cout << destinations[i] << "\n";
20    }
21
22    std::vector<std::string>::const_iterator c;
23    for (c=destinations.begin(); c!=destinations.end(); c++)
24    {
25      std::cout << *c << "\n";
26    }
27
28    destinations.insert(destinations.begin(), "Sydney");
29    destinations.insert(destinations.begin()+1, "Moscow");
30    destinations.push_back("Frankfurt");
31    std::cout << "Length of vector is "
32            << destinations.size() << "\n";
33    std::cout << "Entries of vector are\n";
34
35    for (c=destinations.begin(); c!=destinations.end(); c++)
36    {
37      std::cout << *c << "\n";
38    }
```

```
39
40    destinations.erase(destinations.begin()+3,
41                       destinations.end());
42    std::cout << "Length of vector is "
43              << destinations.size() << "\n";
44    std::cout << "Entries of vector are\n";
45
46    for (c=destinations.begin(); c!=destinations.end(); c++)
47    {
48       std::cout << *c << "\n";
49    }
50
51    sort(destinations.begin(), destinations.end());
52    std::cout << "Length of vector is "
53              << destinations.size() << "\n";
54    std::cout << "Entries of vector are\n";
55
56    for (c=destinations.begin(); c!=destinations.end(); c++)
57    {
58       std::cout << *c << "\n";
59    }
60
61    return 0;
62 }
```

8.3.2 Sets

A set is an STL container where new entries are only stored if they are distinct from the entries already stored. The machinery for maintaining the distinctness of the entries is *abstracted* from the user. One might implement a set as an unordered list of elements, so that each insertion requires a membership test that may involve an equality check with all elements of the existing set. One might make a more efficient implementation using an *ordered* list, so that membership tests involve fewer equality checks against existing members. The STL set actually uses a more efficient structure[1] so that it is able to guarantee the efficiency of all possible set operations. It is only possible to make an efficient set implementation if the elements of the set can be ordered. We will demonstrate the set container by using the class of points in two dimensions whose members have coordinates that take integer values. As the items in a set have to be comparable, we need to define an ordering on points in two dimensions, which we do by overloading the "less than" operator for these points. If we are comparing two points P_0 and P_1, which represent the points (x_0, y_0) and (x_1, y_1), we say that $P_0 < P_1$ if $x_0 < x_1$, and $P_0 > P_1$ if $x_0 > x_1$. Only if $x_0 = x_1$ we say that $P_0 < P_1$

---

[1] The STL set is implemented as a tree structure known as a *red-black search tree*.

if $y_0 < y_1$, and $P_0 > P_1$ if $y_0 > y_1$. If $x_0 = x_1$ and $y_0 = y_1$ then the points P_0 and P_1 are identical: the set would only store one instance of these two.

The class `Point2d` representing the class of points in two dimensions is given in the listing below. This class has two member variables, x and y, that store the x- and y-coordinates. There is also a constructor that allows us to initialise the coordinates, and an overloaded "less than" < operator that allows us to order points in two dimensions as described above.

```
1   class Point2d
2   {
3   public:
4       int x, y;
5       Point2d(int a, int b)
6       {
7           x = a;
8           y = b;
9       }
10      bool operator<(const Point2d& other) const
11      {
12          if (x < other.x)
13          {
14              return true;
15          }
16          else if (x > other.x)
17          {
18              return false;
19          }
20          else if (y < other.y)
21          {
22              // x == other.x
23              return true;
24          }
25          else
26          {
27              // x == other.x and
28              // y >= other.y
29              return false;
30          }
31      }
32  };
```

In the listing below, we create a set of instances of the class `Point2d`. When using the set container, we must include the `set` header file (line 1). In line 7 we create a set, made up of instances of the class `Point2d`, that is called `points`. In lines 9–12, we attempt to insert four points into this set using the `insert` method associated with sets. Two of these points—the origin and the point (0, 0)—are identical, and so only one is stored. This is seen in lines 14 and 15 where we print out the size of the set, which is 3. Note how the iterator may be used in lines 17–21 of the code to print the member variables of the class of points in line 20.

```
 1  #include <set>
 2  #include <iostream>
 3  #include "Point2d.hpp"
 4
 5  int main(int argc, char* argv[])
 6  {
 7      std::set<Point2d> points;
 8      Point2d origin(0, 0);
 9      points.insert(origin);
10      points.insert(Point2d(-2, 1));
11      points.insert(Point2d(-2, -5));
12      points.insert(Point2d(0, 0));
13
14      std::cout << "Number of points in set = "
15                << points.size() << "\n";
16
17      std::set<Point2d>::const_iterator c;
18      for (c=points.begin(); c!=points.end(); c++)
19      {
20          std::cout << c->x << " " << c->y << "\n";
21      }
22
23      return 0;
24  }
```

8.4   A Survey of Some New Functionality in Modern C++

At the close of Chap. 4 we thought it pertinent to give you an indication that some features of C++ have moved on since the first edition of this book was written. In Sect. 4.4 we introduced two of the new smart pointer constructs which have been implemented in compilers that conform to modern C++ standards. This enabled us to indicate that, whereas in former days all dynamically allocated memory was the responsibility of the programmer, there are now ways to ensure that certain pointers are not aliased (via the unique_ptr type) and to automatically garbage collect certain variables (via shared_ptr).

Now, towards the close of this chapter, we would like to introduce a selection of some of the other features available in modern C++ standards such as C++11. With the exception of smart pointers, we have deferred writing about any of these new features until now because most of the features are *templated* over a type. By introducing modern C++ features here we only intend to scratch the surface. As in Sect. 8.3, this section is intended only as a brief survey of some of the available features. We have deliberately selected those features which we have found most helpful in the years since we wrote the first edition of this book—in the belief that these features will prove useful to other computational scientists.

Note that with a current version of the GNU C++ compiler, all code fragments in this section require that the compiler is explicitly told that it is compiling code that conforms to a newer standard of C++ than its default. This may mean invoking

```
g++ -std=c++11 -o TestingCode TestingCode.cpp
```

or something similar on the command-line.

8.4.1 The `auto` Type

After reading Sect. 8.3 you may have been left believing that templates are all double colons and angle brackets. Worse, that whenever you want to iterate over a vector or set which you have created, then you will need to remember the exact form of the iterator type. The good news is that much of the writing of these types can now be simplified via *automatic type inference*. This not only saves on writing, but it also makes templated code more readable, by removing some of the lengthy type names. This relies on one simple rule.

Rule: if the type of a new variable can be inferred by the compiler at the point of its initialisation then the type may be replaced by `auto`.

For example, in the code fragment below, there is enough information for the compiler to infer that `i` ought to be an integer variable: at its initialisation it is given the value 1 (an integer). Meanwhile the variable `x` which is initialised to a floating point value is given the inferred type `double`. Note that each of these two lines contain both the `auto` type and an assignment. Neither of these lines can itself be split across two lines, because if the type is separated from the initialisation then the type can no longer be inferred.

```
1    // Requires C++11 or above
2    auto i = 1;
3    auto x = 22.5;
```

It is worth pointing out that the full power of automatic type inference is not demonstrated by the above example and, furthermore, that inferring simple types as in this example is potentially dangerous because the programmer may find unexpected behaviour if the compiler infers a different type to the type presumed by the programmer. For example, the code below will print to console that `x` contains the value 22 which may not be what the programmer intended. This is because, on line 1, the compiler will infer that the type of `x` ought to be `int`. A programmer, reading line 2, may assume that `x` ought to be of type `double`, but at this point it's too late—the type is fixed. The programmer could repair this issue either by initialising the value of `x` to 22.0 or by using `double` as the explicit type for `x`.

```
1   auto x = 20;  // Compiler infers x as int
2   x += 2.5;     // Programmer might assume x is double
3   std::cout << "x = "<<x<<"\n"; // x = 22
```

The real power of the `auto` keyword comes in places where the onus used to be on the programmer to write out a lengthy type name. For example in Listing 8.3, loops beginning at lines 23, 35, 46 and 56 all rely on a `const_iterator` which is declared once globally in line 22. This was largely done to keep the code compact. Note though that the style of globally declaring a loop iterator in Listing 8.3 is in direct contravention to point 4 in our tips on coding style (given in Sect. 6.6). In the code below we have indicated how the code in any of these `for` loops over the vector `destinations` may be replaced concisely with one which has a locally-declared iterator of type `auto`.

```
8    std::vector<std::string> destinations;
9    // (Fill the vector with names...)
10
11   for (auto c=destinations.begin();
12             c!=destinations.end(); c++)
13   {
14      std::cout << *c << "\n";
15   }
```

8.4.2   Some Useful Container Types with Unified Functionality

Modern C++ provides `std::array` which is a useful replacement for the small size, statically allocated array (as found in Sect. 1.4.5). The idea behind this array type is to provide a uniform way to access and use arrays. It is templated by the type of object it contains and its size. In terms of access to its elements it can behave exactly like the old style plain array: the element index in square brackets is used to read or write individual elements. This is demonstrated below where an old-style array and a new-style array are created on lines 1 and 2 respectively. In the assert statement on line 3, elements of the two arrays are compared using the same syntax.

```
1   int odd[4] = {1,3,5,7};              // Old style
2   std::array<int, 4> even = {2,4,6,8}; // New style
3   assert( odd[3] + 1 == even[3] );
```

However, there are two main ways in which the new `std::array` is very different to the old array. In both of these respects it behaves a lot more like a fixed-size version of `std::vector`. The first difference is that many of the vector functions, such as begin() and size(), are available in the templated array class. The second difference is that arrays can now be passed into functions as first-class objects. That is, whereas old-style arrays are always sent to functions as pointers (a fact we exploited in Sect. 5.2.4) the new type can either be copied or sent as a reference.

Because the new-style array is built around the same infrastructure as the previous STL structures it interacts with many of them in the way one might expect. Many structures may be initialised using the *initialiser list* style (a list of elements in curly braces). One can also convert between many of the structures by copying data between objects.

In the code below, after initialising an array in line 1, we then copy the array contents into both a vector and a set in lines 2–4. Note that the syntax for the two operations is that same but that the data are converted to different underlying representations. Finally the representation is tested in line 5, where we expect that the set, which has no duplicates, will contain fewer members.

```
1   std::array<int, 4> num_array = {1, 3, 5, 3};
2   std::vector<int> num_vector(num_array.begin(),
3                               num_array.end());
4   std::set<int> num_set(num_array.begin(), num_array.end());
5   assert( num_set.size() < num_vector.size() );
```

Modern C++ also contains a light-weight mixed-type *tuple*. This allows us to put pieces of data together in one place in a modular way, so that it is similar to a small class with no methods. The tuple is a generalisation of the existing STL data structure pair (which was restricted to having exactly two pieces of data). An example of the use of a tuple is given below. The new tuple explorer is required to represent information about a book via two strings and a number. It is clear from lines 4 and 5 that access to the member data in the tuple is possible (though the syntax may look a little strange). Finally in line 7 we tidy up some of this new strange syntax by using the keyword auto which enables us to initialise a reasonably complicated mixed-type tuple in a single line.

```
1   std::tuple<std::string, std::string, int> explorer =
2               std::make_tuple("The explorer",
3           "Katherine Rundell", 2017);
4   std::cout<<"Title is "<<std::get<0>(explorer)<<"\n";
5   std::cout<<"Published: "<<std::get<2>(explorer)<<"\n";
6
7   auto h=std::make_tuple("The hobbit", "JRR Tolkien", 1937);
```

8.4.3   Range-based for Loops

A very useful feature of modern C++ is the range-based loop. This is sometimes known in other languages as a "for each" loop but, as we will see later, "for each" has a reserved meaning in modern C++. The range-based loop provides the programmer with a way of iterating over each and every member of a particular container (array or vector, for example) without having to worry about how many members there are, or about the exact mechanism of iteration.

The most simple way to demonstrate this is with an iteration over an intialiser list. Here the variable even, which is local to the for loop takes on all the values in the given list up to, and including the value 8:

```
1   for (int even : {2,4,6,8})
2   {
3       std::cout << even << "\n";
4   }
```

The range-based loop is available for all structures which might be iterated over: arrays, vectors, sets, maps and so on. In each case the meaning of the range-based loop is to iterate over the container in the same way that the container's regular iterator might behave (but with a far more compact syntax). If we use a range-based loop on a std::set then we expect to see each element of the set exactly once, but with no guarantee on the order in which they appear. If we use a range-based loop on a std::vector then we will see each item according to their position in the vector.

We can now re-visit the example code for the STL vector type in Listing 8.3 and again re-write those loops which are iterating over the members of the vector and printing them out. Note that in the code below we have taken advantage of the array initialiser in lines 8–9 in order to rapidly fill the vector destinations with content.

In lines 11–15 we show the normal usage for a range-based loop over a vector. On each iteration of this loop the variable city, which is local to the loop, is assigned a value which is a *copy* of an item in the vector. This means that the content of the underlying vector cannot be changed: any modifications to the copied string in the variable city will stay local to the loop. If, on the other hand, we intended to change the contents of destinations, then we would do so by using a reference to the items. The use of a reference in a range-based loop is demonstrated in lines 16–21. Here each of the city names in the vector is modified using simple string concatenation. Finally, in line 23, we show that a combination of a range-base loop, the auto keyword, and writing the loop without braces leads to a highly compact way to express the same code. The loop in line 23 is equivalent to that in lines 11–15. If we wanted to make modifications to the vector then we could insist that the local variable were a reference by writing auto& instead of auto for the type name.

```
8    std::vector<std::string> destinations =
9         {"Paris", "New York", "Singapore"};
10
11   // Range-based loop
12   for (std::string city : destinations)
13   {
14     std::cout << city << "\n";
15   }
16   // Use a reference to alter the members
17   for (std::string& r_city : destinations)
18   {
19     r_city = r_city + " (modified)";
20     std::cout << r_city << "\n";
21   }
22   // A very compact form
23   for (auto city:destinations) std::cout<<city<<"\n";
```

8.4.4 Mapping Lambda Functions

We close this survey with the "for each" function, which is intended to take a function
and apply it to every member of a container (for example a vector). This type of
functionality is called a "map" in some languages. It works by taking as arguments
the beginning and end of a range to be iterated over, and the function that should
be applied. In the most straightforward form the "function" might just be the name
of function which has been defined elsewhere, but it becomes more powerful when
the function can be declared locally: inside the current scope, or even within the
for_each function itself. The local declaration of a function is known as a *lambda
closure* by computer scientists.

In the following code fragment, we apply functions which double each element
of a vector. The first time this happens is on line 7 where the function name twice
appears in the third argument. Now this twice function might have been declared
externally (as indicated by the comment on line 4) but, instead, it is declared on line
6. Square brackets here indicate that what follows is a function, with round brackets
around the argument, braces around the function body, and a final semicolon. We
have used auto for the type of twice because its real type is a function from int&
to void. Lines 10–11 show that the function does not need a name. Instead we can
just declare the form and definition of the function in place. This is a very compact
form, but perhaps renders the code less readable.

```
1    std::vector<int> evens = {2,4,6,8};
2
3    // Locally declare the equivalent of
4    //   void twice(int& n){ n*=2; }
5
6    auto twice = [](int& n){n *= 2;};
```

```
 7    std::for_each(evens.begin(), evens.end(), twice);
 8
 9    // Compact form
10    std::for_each(evens.begin(), evens.end(),
11                      [](int& n){n *= 2;} );
```

8.5   Tips: Template Compilation

In Sect. 8.1 we presented a templated class `DoubleVector` in which the size of the vector is specified at compile time. Since the size of the vector in `UseDouble-Vector.cpp` (Listing 8.2) is known at compile time, the memory allocation is static.

When building a program to use a templated class such as `DoubleVector` we might follow the pattern laid down in Sect. 6.2.4.1 of placing the *definition* of the class in the file `DoubleVector.hpp` and the *implementation* of the class in the file `DoubleVector.cpp`. We would write a main program to test it and write the rules for compilation into a `Makefile`. There is an unfortunate snag with this plan, because when we instantiate a vector (`DoubleVector<5>`, say) in our main program and compile it, the compiler has no access to the implementation from `DoubleVector.cpp`. The compiler needs to compile code from the `DoubleVector.cpp` file, in which all the instances of DIM are replaced by "5".

There are three strategies which can be used to overcome this *template instantiation* problem.

1. Each file which uses the class may include the implementation of the entire class through the use of `#include "DoubleVector.cpp"`. This means the code compilation may be slower since the entire class must be compiled every time it is used. It also means that care must be taken to ensure that the file `DoubleVector.cpp` is included at most once. (The `#define` mechanism introduced in Sect. 6.2.2 may be suitably adapted for this purpose.)
2. A similar solution is to place the entire class in the file `DoubleVector.hpp`, as we did for `DoubleVector` in Listing 8.1 of Sect. 8.1. This, again, has the disadvantage that the entire class must be compiled every time it is used.
3. A more advanced solution to the problem is *explicit instantiation*. If it is known that we only use `DoubleVector` with a small set of sizes, then we can force the compiler to produce exactly the ones which are needed as it compiles `DoubleVector.cpp` into the object file `DoubleVector.o`. This is done by making an unnamed instance of the class of each required size in the file `DoubleVector.cpp`, as the code fragment below illustrates.

```
1    #include "DoubleVector.hpp"
2
3    template class DoubleVector<5>;
4    template class DoubleVector<7>;
```

8.6 Exercises

8.1 The probability of rain for each of the next N days is to be stored in a double precision floating point array of size N. As the entries of this array are probabilities they should all take values between 0 and 1 inclusive. However, as they have been calculated using a numerical algorithm, these probabilities are only correct to within an absolute error of 10^{-6}: that is, in reality these numbers may be between -10^{-6} and $1 + 10^{-6}$ inclusive. Using the ideas presented in Sect. 8.1, use templates so that when accessing an individual entry of the array:
1. the value stored by the array is returned if it is between 0 and 1 inclusive;
2. the value 0 is returned if the value stored is between -10^{-6} and 0 inclusive;
3. the value 1 is returned if the value stored is between 1 and $1 + 10^{-6}$ inclusive; and
4. an assertion is tripped otherwise.

8.2 Use templates to write a single function that may be used to calculate the absolute value of an integer or a double precision floating point number.

8.3 Use the class of complex numbers given in Sect. 6.4 to create an STL vector of complex numbers. Investigate the functionality of the STL demonstrated in Sect. 8.3.1 using this vector of complex numbers. Note that when you add an object to an STL vector it is a *copy* which is added, so it is imperative that the copy constructor is working as expected.

8.4 Modify the example of an STL set given in Sect. 8.3.2 so that the coordinates of the point are now given by double precision floating point variables. You will now need to think a bit more carefully about what it means for two coordinates to be equal: see the tip on comparing two floating point numbers given in Sect. 2.6.5.

8.5 Use the container std::map<std::string, int> (a mapping from keys of type string to values of type int) to represent a phone book. If you have access to a compiler which is compatible with C++11 or higher then you might consider some of the following ideas.
1. Use an *initialiser list* to populate the phone book with a small list of name-number pairs. See Sect. 8.4.2 for examples of structures initialised in this way, but be aware that a *map* needs to be initialised with a list of lists.
2. Write a for loop to iterate over the contents of the phone book and output all name-number pairs. Try this with the range-based loop that was introduced in Sect. 8.4.3.
3. Write out the entire contents of the phone book using a std::for_each loop and a lambda function, in a similar manner to the loops shown in Sect. 8.4.4.
4. Write functionality to get all names from the map and store them in a vector.
5. Write functionality to get all the numbers from the map into a vector. Then use std::set to detect whether two or more people share the same number.

6. Write the "reverse" map to look up a name when given a number. If you have two people who share the same number then you may find that `std::multimap` is useful.
7. Re-write the map so that, instead of each name mapping to a single number value, it maps to a `std::tuple` consisting of a number and an email address. See Sect. 8.4.2 for example use of *tuples*.

Errors, Exceptions and Testing

In Sect. 1.6 we introduced the concept of an assert statement. This is a way of forcing your program to terminate execution, should something unexpected happen. The program which motivated the use of assertion in Sect. 1.6 was one which calculated the square root of a number entered at the command-line. Here is a version of that program where the assertion has been removed by turning it into a comment.

```
1   #include <iostream>
2   #include <cassert>
3   #include <cmath>
4
5   int main(int argc, char* argv[])
6   {
7       double a;
8       std::cout << "Enter a non-negative number\n";
9       std::cin >> a;
10      //Run without assertion: assert(a >= 0.0);
11      std::cout << "The square root of "<< a;
12      std::cout << " is " << sqrt(a) << "\n";
13      return 0;
14  }
```

What happens when a user ignores the request and enters a *negative* number at the command line? Without the assert statement on line 10 it is likely that the program will complete without error. This is because the computer's floating point unit renders the result of some calculations such as sqrt(-1.0) as "not a number" or nan for short.

© Springer International Publishing AG, part of Springer Nature 2017
J. Pitt-Francis and J. Whiteley, *Guide to Scientific Computing*
in C++, Undergraduate Topics in Computer Science,
https://doi.org/10.1007/978-3-319-73132-2_9

```
Enter a non-negative number
-1
The square root of -1 is -nan
```

Other examples of floating point operations which produce the answer nan include $0.0/0.0$ and $\log(0.0)$. Some calculations such as $1.0/0.0$ will resolve to a floating point representation of infinity (inf). In a scientific program, once one variable has been set to nan or inf then this value is likely to propagate to later parts of the calculation. It is normally best to check for this sort of error at the earliest possible stage so that computation is not wasted. In this context, it would be prudent to check in any piece of code that uses division, square root, logarithms etc. that the values of all the arguments are in a sensible range. As we have already seen, assertions are one method of checking such arguments. In this chapter, we will see that exceptions are another method of checking that are more flexible in some ways. We will also introduce techniques for testing software, to allow software to be developed in a sustainable manner with as few errors as possible.

9.1 Preconditions

Every section of a program (where a "section" could be a function, method, block, for-loop iteration body etc.) can be thought of as having the task to produce a *postcondition* when given a valid *precondition*. The postcondition of the program above (the thing which it is tasked to do) is that it prints the square root of a given number. It does this subject to the precondition that the number is nonnegative.

Consider a method which finds all the roots of a function $f(x)$ in the half-open range $x_{min} \le x < x_{max}$. This method might need to assume as a precondition that the function f is continuous and differentiable over the same range $x_{min} \le x < x_{max}$. More trivially, it might also need to assume that $x_{min} < x_{max}$. What should happen if $x_{min} > x_{max}$ or $x_{min} = x_{max}$? If the precondition for correct functionality is not met then what should happen? Before we answer this question, we will first consider a specific case.

9.1.1 Example: Two Implementations of a Graphics Function

In a particular graphics library, there is a function for rendering a 2-D annulus. This function takes four input arguments: the inner radius, the outer radius and the number of radial and axial segments. The specification of the library says that the outer radius must be bigger than the inner radius and both should be nonnegative. It also says that the segment numbers must be strictly positive. The specification further says that it is valid to give the inner radius as zero, in which case the annulus will be rendered as a 2-D disk with no hole.

There is a cautionary story about a professor who wrote a program for his students which used this graphics function to draw disks. He misread the specification and set the radius values the wrong way round so that the outer value was 0.0 and the inner value was 1.0. Without realising his mistake, he distributed the program source code to his students, some of whom began to complain that it would not run.

The problem was that the students whose code would not run were using a different implementation of the library. The two different implementations of the same specification were dealing with errors in different ways. The implementation of this function in the graphics library as used by the professor contained a check for his type of error which silently fixed the problem by interchanging values in a manner similar to the code given below.

```
1   #include <cassert>
2   void RenderAnnulus(double innerRadius, double outerRadius,
3                      int slices, int segments)
4   {
5       //A "helpful" implementation fixes the input
6       //so RenderAnnulus(1.0, 0.0, 30, 3); will work
7       if (innerRadius > outerRadius)
8       {
9           //The arguments are the wrong way round
10          //Swap them
11          double temp = innerRadius;
12          innerRadius = outerRadius;
13          outerRadius = temp;
14      }
15      //...then render the annulus
16  }
```

Meanwhile, the students who complained that the program was not running properly were using a library implementation in which the annulus function terminated on reaching this type of error. The listing below shows that this termination behaviour can easily be implemented by checking the precondition with an assertion.

```
1   #include <cassert>
2   void RenderAnnulus(double innerRadius, double outerRadius,
3                      int slices, int segments)
4   {
5       //Another implementation only checks the input
6       //so RenderAnnulus(1.0, 0.0, 30, 3); trips an assertion
7       assert (innerRadius < outerRadius);
8       //...then render the annulus
9   }
```

The "helpful" implementation, as used by the professor, was in reality making a bug in his code invisible—only for it to become embarrassingly visible in the other implementation. Both implementations are *correct* in the sense that they follow the specification and perform the correct operations provided that the preconditions are met. Unfortunately, the library specification left the handling this kind of error open to interpretation.

9.2 Three Levels of Errors

Some of the most important decisions that a programmer has to make are about how errors should be treated. What should happen if the user misreads a prompt and enters some invalid input? What should happen if the application writer accidentally permutes the input arguments of a library function? What should happen if some numerical scheme has generated inf or nan because of divergence?

The answer to all these questions is the same: "It depends". It's good to treat errors differently depending on their severity, both in terms of how likely they are to happen and in terms of how easy it might be to fix the problem and carry on. The difficult balance of knowing how severe an error might be is illustrated by the RenderAnnulus story in Sect. 9.1.1 where the programmers of different library implementations chose to deal with the same error in completely different ways. One set of programmers decided the error was trivial to fix, while the other set decided to abort the program.

We propose a strategy for handling errors which is built on a framework of three levels of errors.

1. If the error can be fixed safely, then *fix* it. If need be, warn the user.
2. If the error could be caused by some reasonable user input then throw an *exception* up to the calling code, since the calling code should have enough context to fix the problem.
3. If the error should not happen under normal circumstances then trip an *assertion*.

These three basic levels could be further refined. You may distinguish between errors that trip assertions (which are normally removed in optimised code) and errors that should halt the program under all circumstances. At the other end of the scale, you might distinguish between error fixes which are silent and those which should warn the user that something has been changed.

The *exception* level of error is a compromise between patching the problem to carry on, and stopping completely. It is used in circumstances where the caller of a function may have enough information to be able to deal with the error. For example, a nonlinear Newton root finder may diverge and hence signal an error, but the programmer may know that the original task in question can still be solved by calling the same function with a different initial guess, or by calling it with a damping factor, or by calling a bisection root finder. The logic would be to first try the Newton solver, but if that function signalled an error then to find the root using a more expensive bisection routine.

9.3 Introducing the Exception

An exception in C++ is a way of interrupting the normal flow of control of a program and *throwing* a bundle of information back to the calling code. This bundle of information is encapsulated inside an object. We define in this section a class called `Exception`, but objects of any class may be thrown between functions to signal an error.

The use of exceptions requires the keywords `try`, `throw` and `catch`.

- `try` is used in the calling code and tells the program to execute some statements in the knowledge that some error might happen.
- `throw` is used when the error is identified. The function called will encapsulate information about the error into an `Exception` object and throw it back to the caller.
- `catch` is used in the calling code to show how to attempt to fix the error. Every block of code that has the `try` keyword must be matched by a `catch` block.
- Exceptions which are not caught by the calling code may cause the program to halt.

When an error occurs we want the code to "throw" two pieces of information: a one-word summary of the problem type and a more lengthy description of the error. We write a class `Exception` (shown below) to store these two pieces of information, and with the ability to print this information when required.

Listing 9.1 Exception.hpp

```
1  #ifndef EXCEPTIONDEF
2  #define EXCEPTIONDEF
3
4  #include <string>
5
6  class Exception
7  {
8  private:
9      std::string mTag, mProblem;
10 public:
11     Exception(std::string tagString, std::string probString);
12     void PrintDebug() const;
13 };
14 #endif //EXCEPTIONDEF
```

Listing 9.2 Exception.cpp

```
1  #include <iostream>
2  #include "Exception.hpp"
3  //Constructor
4  Exception::Exception(std::string tagString,
5                       std::string probString)
```

```
6    {
7        mTag = tagString;
8        mProblem = probString;
9    }
10
11   void Exception::PrintDebug() const
12   {
13       std::cerr << "**  Error ("<<mTag<<") **\n";
14       std::cerr << "Problem: " << mProblem << "\n\n";
15   }
```

9.4   Using Exceptions

In Listing 3.4, we read from a named file Output.dat. We assumed that this file
existed and tripped an assertion if it did not. In the code below, we present a more
sophisticated program for opening a file which uses exceptions to attempt to fix the
problem. If the file cannot be opened by the ReadFile function, an exception is
thrown. This is caught by code that prompts the user to enter an alternative file name.
Note that ReadFile takes the name of the file as a C++ string which is converted
to a C string on line 8 (using c_str which was introduced in Sect. 1.4.8).

```
1    #include <iostream>
2    #include <fstream>
3    #include "Exception.hpp"
4
5    void ReadFile(const std::string& fileName, double x[],
6                  double y[])
7    {
8        std::ifstream read_file(fileName.c_str());
9        if (read_file.is_open() == false)
10       {
11           throw (Exception("FILE", "File can't be opened"));
12       }
13       for (int i=0; i<6; i++)
14       {
15           read_file >> x[i] >> y[i];
16       }
17       read_file.close();
18
19       std::cout << fileName <<" read successfully\n";
20   }
21
22   int main(int argc, char* argv[])
23   {
24       double x[6], y[6];
25       try
```

```
26    {
27        ReadFile("Output.dat", x, y);
28    }
29    catch (Exception& error)
30    {
31        error.PrintDebug();
32        std::cout << "Couldn't open Output.dat\n";
33        std::cout << "Give alternative location\n";
34        std::string file_name;
35        std::cin >> file_name;
36        ReadFile(file_name, x, y);
37    }
38 }
```

9.5 Testing Software

It is often the case that you need to take a program which has been developed in the past and seek to extend its functionality, perhaps to address some new research question. Assuming that you are able to understand the working of the original code because it is well-documented (as suggested in the tips given in Sect. 5.10) and has a literate coding style (as suggested in the tips given in Sect. 6.6), there is still a potential pitfall. Suppose you add the new functionality, use it to solve your new research problem, but later discover that the original functionality of the code has changed. Perhaps you are no longer able to reproduce the results which are needed for a publication. This pitfall may have been avoided had an appropriate software testing strategy been used for the original code.

For reasons including those given above, it is universally accepted that software should always be tested to give confidence in the output when a code is executed. There is, however, less agreement on how much effort should be put into testing, and on the methodology to be used for testing software. One reason for the absence of a unified view is that the rigour required depends on many characteristics of the software which we now explain with the aid of examples.

Suppose we have a file that contains many 2×2 matrices that are believed to represent rotations, that has been generated from a piece of software. If $\mathbf{Q}$ is one of these matrices, then $\mathbf{Q}$ must be an orthogonal matrix and so we must have $\mathbf{Q}\mathbf{Q}^{\top} = \mathbf{I}$, where $\mathbf{I}$ is the 2×2 identity matrix, and $\det(\mathbf{Q}) = 1$. Suppose further that a colleague wants to use this file, provided he or she can be reasonably certain that the matrices are indeed orthogonal. If we were to allow this colleague to use this file of matrices then we should first check that the matrices really are orthogonal. We may check this by writing a short program that reads these matrices in and checks that they are orthogonal (subject to rounding errors) by printing to screen any warnings that a matrix isn't orthogonal. In this case it can be argued that this rudimentary method for testing the software is appropriate, as we are checking that the file that we share with our colleague does indeed contain matrices that represent rotations.

Nevertheless we should be aware of the limitations of testing software in such an unsophisticated manner. This method does not ensure that the original software is error free; all we have done is to confirm that the given file does indeed contain orthogonal matrices. For example, if the file is believed to represent 1000 distinct matrices we have not checked that there really are 1000 matrices, or that they are distinct—we may only have 500 distinct matrices, or we may have one matrix that has been printed 1000 times. It is also possible that an error exists in the software used to generate the matrices, and that a subsequent execution of the software generates some matrices that are not orthogonal. Many other potential sources of error also exist.

Consider, by contrast, a piece of software containing many lines of code that controls a mechanical ventilator in the clinical setting. It is clearly of critical importance that as many errors as possible are eradicated from the software before it is used, and so much more rigorous testing of the code is required. Furthermore, it is likely that the software may be updated for future generations of ventilators. It is surprisingly easy to break the original functionality of software when making what appears to be a small extension. It is therefore extremely useful to be able to test the whole code after making even a small modification to this code. The basic technique of testing software described above for the file of matrices is not appropriate in this case, and more sophisticated techniques should be used. The testing of *safety-critical software systems* is a research topic in its own right.

The two examples above illustrate that the effort that should be dedicated to testing software depends on many factors. The first, simpler case required nothing more than a short, disposable C++ program that may easily be written by a competent programmer and requires no more discussion. The second case requires far more attention to the testing strategy. We will now describe some common testing strategies.

9.5.1 Unit Testing

An effective technique for testing software, that is particularly useful for software that may be extended in the future, is known as *unit testing*. When using this technique, a collection of tests are written, known as *unit tests*. Each unit test is designed to test a particular section of the code, for example a single method of a class. Each test should then be executed when new functionality is added; should a test fail then it is clear that the new functionality has broken an existing part of the original functionality.

Unit testing is particularly effective when: (i) each unit test covers only a very small number of lines of the original code; and (ii) each line of the original code is covered by at least one test. Whenever we add a small amount of new functionality we can then re-run each test. If we have broken any existing functionality at least one test would hopefully fail (as each line of code is covered by at least one test). Furthermore, as each test covers only a few lines of code, knowing which tests had failed should help us pinpoint the lines of code where the original functionality had been broken.

In the previous paragraph we explained that, when using unit testing, should existing functionality be broken then at least one test will "hopefully" fail. The reader may expect that, rather than one test hopefully failing, at least one test would *definitely* fail. Unfortunately this assumes that we fully understand the algorithm being used by the software, and have written our unit tests to cover every possible cause of this algorithm failing. For effective unit testing, all possible scenarios must be tested. Suppose, for example, we are writing a graphics application. As part of this application we may want to know where two lines in the (x, y)-plane intersect. This can easily be done by solving two simultaneous equations to calculate the coordinates of the points where the lines meet. We should obviously write a test to check that these coordinates are accurately calculated for two example lines with a unique point of intersection. Despite having written a test that has passed in the example case, there are possibilities where this method does not behave as expected. First, suppose the two lines are identical. They will then intersect at every point. A method written to calculate the intersection of these lines will either fail, or will return one point on the line. A second case is when the lines are parallel, but don't intersect. Any method used to calculate the intersection of these lines would not be able to give a correct answer. To fully test this code we should write tests that cover all possibilities highlighted here. If we don't do this then it is possible that the errors described here may occur, and will propagate into other parts of the code. This may cause other tests to fail, identifying that a problem exists. However the cause of the failing test will not be as clearly located, and may require many tedious and frustrating hours of debugging to pinpoint. We therefore encourage programmers to write tests that cover all possible scenarios.

One highly recommended strategy for writing unit tests is to write the tests for new functionality *before* adding this new functionality. This test will clearly fail initially. All tests—including the new test—are then run when the new functionality has been added, ensuring that both the new functionality has been correctly implemented and that the existing software has not been broken. This method of software development is known as *test driven development*.

Several C++ testing framework libraries exist, such as `CxxTest`, `Boost.Test` and `googletest`. These are designed to help you structure your testing, and we recommend using one of these libraries when writing a suite of tests.

9.5.2 Extending Software

It is very rare that a software package is written from scratch. It is more common for existing software to be extended. For example, you may be expected to extend the functionality of software written by a colleague. Alternatively you may develop software that is underpinned by libraries from external sources. Even if the existing software is believed to be reliable, the user should at least test their own implementation of the functionality offered. This can be done by simply testing all functionality of the software, without understanding the implementation of the functionality—this is known as *black box testing*. This is appropriate for well-supported, mature libraries, that are widely accepted to be robust and reliable. There are, however, potential pitfalls associated with black box testing. Suppose we are using some externally written software that contains the functionality to solve a linear system. When using black box testing, we would simply check that this functionality works for a given linear system. However, if we had taken a course in linear algebra, we would know that there is no solution to some linear systems, and a non-unique solution to other linear systems. To limit errors from the externally written software propagating into the code we develop, we may want to know how the software handles these cases; this will depend on the implementation of the functionality for these special cases. In these cases we would deliberately test the externally written software by choosing one example linear system with no solution, and one example linear system with a non-unique solution. In this case we may also investigate the algorithm that underpins the functionality of the system, allowing us to understand how the given software handles these systems of equations. This variety of testing is known as *white box testing*.

We now explain how both black box testing, white box testing and test driven development may be carried out. We illustrate the concepts discussed above using the CxxTest library, applied to the class of complex numbers developed in Sect. 6.4. We focus on the principles of testing, thus allowing the reader to apply these principles to other testing libraries. As such, we do not focus heavily on the details of using CxxTest; a user guide for this library may be found at http://www.cxxtest.com.

9.5.3 Black Box Testing

We illustrate black box testing using the class of complex numbers developed in Sect. 6.4. As explained earlier, when using black box testing we check that the functionality works correctly without inspecting the implementation. In Listing 9.3 we have written a suite of tests for some of the public methods contained in the header file for this class (given in Listing 6.9); we leave the remainder of the black box testing of these public methods as an exercise. As explained earlier, we use the C++ testing framework library CxxTest for writing these tests. We reiterate that we are focusing on how suitable tests may be written, rather than explaining how to use the CxxTest library. Nevertheless, a few comments on this library are necessary to allow the reader to understand the tests written. First, lines 5–6 and 78 may be considered to be a wrapper that allows us to use the functionality of this library (after it has been installed). Within this wrapper we have written a collection of unit tests: Test-DefaultConstructor (line 8); TestCustomisedConstructor (line 17); TestCalculatePower (line 36); and TestAgainstStdLibrary (line 61). Within these tests, we test that a floating point variable resulting from a calculation is equal to the true value, subject to ignoring the effects of rounding errors as described in Sect. 2.6.5. If, for example, we were using assertions to check that two double precision variables x and y differed by less than some value epsilon, we would write

```
assert( x-y < epsilon && x-y > -epsilon);
```

or, slightly more compactly

```
assert(fabs(x-y) < epsilon);
```

To write this as a test using the CxxTest library, rather than an assert statement, we would use the specially defined CxxTest assertion

```
TS_ASSERT_DELTA(x, y, epsilon);
```

which, rather than acting as an assertion, would simply report a failure if x and y differ by at least some value epsilon. Many other test assertions are offered by the CxxTest library. When the tests have been written, the library may then be used to generate a test runner that may be compiled so that the tests may be executed. This executable would then report which tests had passed, and which tests had failed. Further details on the features available, and instructions on how to install and use these libraries may be found at http://www.cxxtest.com. We now explain why the tests given in Listing 9.3 are suitable for black box testing of both constructors, and the members CalculateModulus, CalculateArgument and Calculate-Power.

We begin by testing the default constructor. This constructor was written with the intention that both the real part and the imaginary part of a complex number created using this constructor should be initialised to zero. A suitable test for this constructor is to check that an instance of a complex number created using this constructor has zero modulus. Clearly this assumes that the method Calculate-Modulus correctly calculates the modulus of this complex number. As such, this test may be considered to also test the method CalculateModulus, albeit with a particularly simple input. This test may be found in lines 8–15 of the listing. Line 8 defines a test called TestDefaultConstructor. Line 12 then defines an instance of a complex number that is created using the default constructor, and line 13 calculates the modulus of this complex number. Finally, in line 14, we use the function TS_ASSERT_DELTA to test that the calculated modulus really is within 10^{-16} of the true value of zero remembering, as discussed in Sect. 2.6.5, that two floating point numbers that should (mathematically) be equal may differ slightly due to rounding errors.

The test between lines 17 and 34 is intended to test the customised constructor. This constructor allows an instance of a complex number to be generated initialising the real and imaginary parts to specified values (lines 21–23). As the real and imaginary parts of the complex number are private members with no methods that allow us to access these members, we may only confirm the real and imaginary parts of the complex number are correctly initialised by confirming that both the modulus (lines 26–28) and the argument (lines 31–33) of the complex number are correct. We note that this test also allows testing of the members CalculateModulus and

CalculateArgument. It is also worth noting that, as we are treating the class as a black box, we have not copied code from the original class and we are instead calculating the modulus and argument via independent means.

Our next test is to test the member CalculatePower (lines 36–60). In this test we use the customised constructor to create a complex number with non-zero real and imaginary parts (lines 40–42), and calculate the modulus and argument of this number (lines 43 and 44). We test CalculatePower by raising the original complex number to the power of 2, and calculating the modulus and argument of this squared complex number (lines 48–50). We then use properties of complex numbers to check that the modulus of the squared complex number is correct (lines 54 and 55) and that the argument of the squared complex number is correct (line 59).

Our final test in this section is to test some of our functionality against a trusted complex number class std::complex (lines 61–77). The C++ library version of $3 - 4i$, std_z, is initialised on line 64. Note that the std::complex is templated with a floating point number type in angle brackets. Here we use double, to match the type of the private data in our own class, but the class also allows for complex numbers with are stored as float. Notice that the syntax of the functions on std_z is completely different to our own. Despite this, the mathematical specification is the same and, consequently, we may perform the same tests on them in tandem.

Listing 9.3 Black box testing of the class of complex numbers

```
1   #include <cmath>
2   #include <cxxtest/TestSuite.h>
3   #include "ComplexNumber.hpp"
4
5   class ComplexNumberTestSuite : public CxxTest::TestSuite
6   {
7   public:
8       void TestDefaultConstructor(void)
9       {
10          // Test default constructor sets complex
11          // number to zero
12          ComplexNumber z;
13          double mod_z = z.CalculateModulus();
14          TS_ASSERT_DELTA(mod_z, 0.0, 1.0e-16);
15      }
16
17      void TestCustomisedConstructor(void)
18      {
19          // Use constructor that allows us to specify
20          // real and imaginary parts of a complex number
21          double real = 4.0;
22          double imaginary = -3.0;
23          ComplexNumber z(real, imaginary);
24
25          // Test that modulus is correct
26          double modulus = z.CalculateModulus();
27          double true_modulus = 5.0; // (3,4,5) triangle
```

```
28        TS_ASSERT_DELTA(modulus, true_modulus, 1.0e-8);
29
30        // Test argument is correct via different function
31        double argument = z.CalculateArgument();
32        double true_argument = -asin(3.0/5.0);
33        TS_ASSERT_DELTA(argument, true_argument, 1.0e-8);
34    }
35
36    void TestCalculatePower(void)
37    {
38        // Specify a complex number, z, and calculate the
39        // modulus and argument
40        double real = 4.0;
41        double imaginary = -3.0;
42        ComplexNumber z(real, imaginary);
43        double modulus_z = z.CalculateModulus();
44        double argument_z = z.CalculateArgument();
45
46        // Calculate z*z and calculate the modulus and
47        // argument of z*z
48        ComplexNumber z_squared = z.CalculatePower(2.0);
49        double mod_z_squared = z_squared.CalculateModulus();
50        double arg_z_squared = z_squared.CalculateArgument();
51
52        // Test that:
53        //   modulus of z*z = (modulus of z)*(modulus of z)
54        TS_ASSERT_DELTA(mod_z_squared, modulus_z*modulus_z,
55                        1.0e-8);
56
57        // Test that:
58        //   argument of z*z = 2*(argument of z)
59        TS_ASSERT_DELTA(arg_z_squared, 2.0*argument_z, 1.0e-8);
60    }
61    void TestAgainstStdLibrary()
62    {
63        ComplexNumber z(4.0, -3.0);
64        std::complex<double> std_z(4.0, -3.0);
65        TS_ASSERT_DELTA(z.CalculateArgument(),
66                        arg(std_z), 1e-8);
67        TS_ASSERT_DELTA(z.CalculateModulus(),
68                        abs(std_z), 1e-8);
69        // Raise both numbers to power 5
70        ComplexNumber z_5=z.CalculatePower(5.0);
71        std::complex<double> std_z_5 = pow(std_z, 5.0);
72        // Check they are the same
73        TS_ASSERT_DELTA(z_5.CalculateArgument(),
74                        arg(std_z_5), 1e-8);
75        TS_ASSERT_DELTA(z_5.CalculateModulus(),
76                        abs(std_z_5), 1e-8);
77    }
78 };
```

Using the black box testing above has given us some confidence that the members of the class of complex numbers that have been tested have been implemented correctly. Note, however, that we have only used arbitrary choices to test these members. Were we to consider the implementation of these members we may discover some cases that could give unexpected results. We now discuss such an instance when describing white box testing.

9.5.4 White Box Testing

In the class of complex numbers, the real part and the imaginary part of an instance of a complex number are both private members of this class. This made it difficult to black box test the customised constructor of this class in Sect. 9.5.3, where we create an instance of the class of complex numbers and simultaneously initialise both the real part and imaginary part to specified values. The difficulty arose because, within the test we wrote, we were unable to access the private members of the class, and were therefore unable to test directly that these had been set to the correct values. Instead, we tested these values were correct indirectly by testing that the modulus and the argument of the complex number were correct. This, however, relies on the public methods used to calculate the modulus and the argument of a complex number being correct. Should the test of the customised constructor fail, we would not know whether the test failed because of an error in the customised constructor, or in one of the methods used to calculate the modulus and the argument of a complex number. This may be avoided by white box testing where, in contrast to black box testing, we inspect the implementation of the functionality offered by the class of complex numbers. We simply make the test suite in Listing 9.3 (which is a class) a friend of the class of complex numbers, allowing us to access—and test for correctness—the real and imaginary parts of a complex number. In Listing 9.4 we have given an example white box style test of the default constructor. This test may be used to replace the original test (lines 8–15 in Listing 9.3) provided that the test suite itself is given access to the private members of the complex number class via "`friend class ComplexNumberTestSuite;`".

Listing 9.4 Extract from white box testing of the class of complex numbers

```
8     void TestDefaultConstructorWhiteBox(void)
9     {
10        // Test default constructor sets to zero.
11        // Add to ComplexNumber.hpp :
12        // friend class ComplexNumberTestSuite;
13        ComplexNumber z;
14        TS_ASSERT_DELTA(z.mRealPart, 0.0, 1.0e-16);
15        TS_ASSERT_DELTA(z.mImaginaryPart, 0.0, 1.0e-16);
16     }
```

A second use of white box testing may be illustrated by creating an instance of the class of complex numbers using the default constructor. This default constructor will set both the real and the imaginary part of this complex number to zero. The modulus of this complex number is clearly zero. However, the argument of this complex number is given by `atan2(0.0,0.0)`. Mathematically, this is arctan(0/0). As 0/0 is not defined, it is not immediately clear what the result of `atan2(0.0,0.0)` is. To find out, we visit the C++ reference page at http://www.cplusplus.com/reference/cmath/atan2/, where we discover that a *domain error* occurs.[1] This is to be avoided, and so we should update the method `CalculateArgument` given in Listing 6.10 to take account of this special case. An appropriate course of action, that is followed by the scientific computing environment MATLAB, is to set the argument of the complex number zero to 0. We leave the implementation, and testing, of this as an exercise.

9.5.5 Test Driven Development

We have already recommended using test driven development to extend software. When using this technique we first write the tests that are required to test the new functionality, forcing us to be very clear about what we expect our modified software to achieve. These new tests will clearly fail initially, as the new functionality does not yet exist. The new functionality will usually first require some refactoring of existing code, for example modifying an existing constructor to take account of extra data that is now associated with a class to implement the new functionality. If you have a well written and maintained suite of tests you can then run these tests to ensure that you haven't broken any existing functionality. The new functionality is then added, and the tests originally written are run to ensure that the new functionality behaves as expected.

For some applications of complex numbers—for example: the calculation of powers of complex numbers; investigation of the stability of a numerical method for solving initial value ordinary differential equations; and integration of complex numbers around poles—it is convenient to have access to the modulus and argument of a complex number. Rather than calculate these quantities every time they are used by using the methods `CalculateModulus` and `CalculateArgument` that already exist within the class of complex numbers, we could modify the class so that the class contains the private members `mModulus` and `mArgument` to represent these quantities. Should we do this, we would then have to decide whether to introduce the members `mModulus` and `mArgument` instead of the existing members `mRealPart` and `mImaginaryPart`, or in addition to these existing members.

If we modify the class of complex numbers so that we include the private members `mRealPart`, `mImaginaryPart`, `mModulus` and `mArgument` we will have to

---

[1] If you were to use the C version of the trigonometry functions, rather than the C++ one, then you will find that `atan2(0.0,0.0)` gives no error and is defined to be 0.

modify other methods in the class so that all of these members are specified whenever an operation is performed on an instance of the class. If we decide to only include the members mModulus and mArgument we will have to modify other methods in the class to specify these members, rather than mRealPart and mImaginaryPart, whenever an operation is performed on an instance of the class. Whatever choice is made, much of the existing functionality of the class will need to be altered. That is, we will have to refactor the code. This illustrates the importance of having a collection of well written unit tests that each cover a small fraction of the whole functionality. Should any of the existing functionality be broken when the code is refactored, at least one test should fail. The location of the error(s) should then be highlighted.

We leave the implementation of this new functionality as an exercise.

9.6 Tips: Writing Appropriate Tests

In this chapter we have attempted to convince you that an appropriate a collection of unit tests will increase the reliability and longevity of your software. These unit tests should each test a very small part of your code, and each line of software should be covered by at least one test. This testing strategy is, however, underpinned by the assumption that the tests are suitable. The following tips may help you to write appropriate tests, and to get the most out of this technique.

1. Use a C++ testing framework library, such as CxxTest, Boost.Test or googletest. This will help you structure your tests.
2. Add one or more tests for every new piece of functionality, no matter how small the added functionality is.
3. Make tests definitive—they should either pass or fail. However, beware of floating point tolerances and allow for rounding errors in calculations.
4. Remember to write tests for *corner cases*. These are test inputs which may be rare, but might cause problems—collinear triangles, singular matrices, the complex number $0 + 0i$ etc.
5. Rather than spreading test input parameters randomly or evenly, it is more efficient to concentrate on the boundary between types of input. For example, if a test input p is supposed to be a probability ($0 \leq p \leq 1$) then check that $p = 1$ gives the correct answer, but that $p = 1.0001$ gives an error.
6. Review your tests from time to time. Add new tests as necessary and remove only those which you know to be redundant.
7. Automate your testing, so that you do not have to remember to run the tests or remember to check the results.

9.7 Exercises

9.1 Extend the `Exception` class given in Listings 9.1 and 9.2 by creating two inherited classes `OutOfRangeException` and `FileNotOpenException`. Each of these two new inherited classes will derive from the `Exception` class in a similar manner to the way the `Ebook` class derived from the `Book` class in Sect. 7.1. The constructors for each of the two classes should take only the `probString` argument to set the `mProblem` member. Each constructor should ensure that the `mTag` member is automatically set in a similar manner to the way the `format` member was set in the constructor of the `Ebook` class. Write a catch block which is able to catch a generic exception but can also differentiate between these two types of error.

9.2 An earlier tip in Sect. 4.3.2 showed how it was possible for bad memory allocation to terminate your program. If you want your program to continue through a memory allocation error there are two ways to cope with the exception: to turn the exception off (and check the value of the pointer) or to catch the exception. Here is some code which demonstrates how to turn off the exception message but still detect bad allocation of memory, without terminating the program.

```
1    double* p_x;
2    p_x = new (std::nothrow) double[1000000000];
3    if (p_x == NULL)
4    {
5        std::cout << "Allocation failed\n";
6    }
7    delete p_x;
```

The proper way to deal with this issue is, of course, to catch the exception. Rewrite the code fragment above so that there is a `try` block around the line of code which attempts to allocate a large vector to `p_x` and demonstrate that you can catch this exception.
[*Hint: The name of the exception class which you need to catch is not* `Exception`. *It is* `std::bad_alloc`.]

9.3 In Exercise 7.3 in Chap. 7, we developed a library for solving initial value ordinary differential equations. Let us suppose that the solution of the ordinary differential equation represents a probability of some event happening as time evolves. The true solution of this equation should therefore be nonnegative, and no greater than one. Of course, due to both rounding errors and errors induced by the numerical approximation used to calculate the numerical solution, this numerical solution may violate these restrictions slightly. In this exercise, we will suggest how to extend the library developed in Sect. 7.3 to handle these requirements in a way that is consistent with the discussion of dealing with errors given in Sect. 9.2.

We will assume that an acceptable value for the absolute error is 10^{-6}. When solving the differential equation, we therefore won't be concerned if the solution for a value of y_i in Exercise 7.3 lies in the interval $-10^{-6} < y_i < 0$. Under these circumstances, we would simply write the value 0.0 to file containing the solution at each time t_i instead of the value y_i. Similarly, if the solution lies in the interval $1 < y_i < 1 + 10^{-6}$ we would write 1.0 to file rather than the value y_i. This is an instance of an error of type #1 in the list given in Sect. 9.2.

Now suppose the value of y_i lies further outside the range of acceptable values than can be attributed to rounding error. The most likely cause of this error is a step size h that is too large. Under these circumstances, an exception should be thrown explaining this. The code that calls the library for solving initial value ordinary differential equations would then know to reduce the step size: a suitable new step size would be half of the step size currently being used. This is an instance of an error of type #2 in the list given in Sect. 9.2.

It is, of course, possible that an error has been made elsewhere in the library or in the code used to call the library. Under these circumstances persisting with making the step size smaller may not solve the problem. We therefore want to terminate the code if the step size h falls below some critical value. This is an instance of an error of type #3 in the list given in Sect. 9.2.

Incorporate the error handling procedure described above into the library for solving initial value ordinary differential equations developed in Exercise 7.3 in Chap. 7. Test this error handling using the example initial value problem

$$\frac{dy}{dt} = -100y,$$

with initial condition $y = 0.8$ when $t = 0$, for the time interval $0 < t < 100$. Investigate how different values of the step size h affect the error handling implemented.

9.4 In Sect. 9.5 we discussed how unit tests could be written for the class of complex numbers developed in Sect. 6.4. In this exercise we will complete the set of unit tests that we started in Sect. 9.5.3.

1. Extend the unit tests given in Listing 9.3 so that all the public methods in the class of complex numbers—listed in the header file given in Listing 6.9—are tested using black box testing.

2. The default constructor for the class of complex numbers initialises both the real and imaginary parts of an instance of a complex number to zero. We noted in Sect. 9.5.4 that the method `CalculateArgument`, as implemented in Listing 6.10, will give a domain error when applied to the complex number zero. By using white box testing, as described in Sect. 9.5.4, we suggested a suitable fix for this problem. Implement this fix, and write a test to ensure you have implemented this fix correctly.

3. Suppose we are writing a piece of software for investigating the stability of a given numerical method for solving an initial value system of ordinary differential equations. This software will require us to evaluate polynomial functions of

a given complex number, and to confirm that the modulus of a complex number is less than unity. We decide to implement this additional functionality by first modifying the class of complex numbers so that the class also contains the private members mModulus and mArgument that represent the modulus and argument of an instance of a complex number. In Sect. 9.5.5 we explained that we could add these members either in addition to the members mRealPart and mImaginaryPart, or instead of these existing members.

In this exercise you should use test-driven development to implement the new functionality. First decide whether or not to include the existing members mRealPart and mImaginaryPart in addition to the new members mModulus and mArgument, and refactor the existing code as necessary. Having done that, introduce new functionality that uses the new members mModulus and mArgument to evaluate polynomial functions of a given complex number, and to determine whether the modulus of a complex number is less than unity.

Developing Classes for Linear Algebra Calculations

In this chapter, we will apply the ideas introduced earlier in this book to develop a collection of classes that allow us to perform linear algebra calculations. We will describe the design of a class of vectors in the body of this chapter. The exercises at the end of the chapter will focus on developing this class further, developing a companion class of matrices, and developing a linear system class that allows us to solve matrix equations.

10.1 Requirements of the Linear Algebra Classes

As explained above, we will develop a class of vectors called `Vector`, a class of matrices called `Matrix` and a linear system class called `LinearSystem`. The vector and matrix classes will include constructors and destructors that handle memory management. These classes will overload the assignment, addition, subtraction and multiplication operators, allowing us to write code such as "`u = A*v;`" where `u` and `v` are vectors, and `A` is a matrix: these overloaded operators will include checks that the vectors and matrices are of the correct size. The square bracket operator will be overloaded for the vector class to provide a check that the index of the array lies within the correct range, and the round bracket operator will be overloaded to allow the entries of the vector or matrix to be accessed using MATLAB style notation, indexing from 1 rather than from zero.

The remainder of this chapter will focus on the development of a class of vectors. The header file for this class is given in Listing 10.1, and the implementation of the methods is given in Listing 10.2. The two variables that each instance of the class are built upon are a pointer to a double precision floating point variable, `mData`, and the size of the array, `mSize`. We have made both of these private members of the

© Springer International Publishing AG, part of Springer Nature 2017
J. Pitt-Francis and J. Whiteley, *Guide to Scientific Computing*
in C++, Undergraduate Topics in Computer Science,
https://doi.org/10.1007/978-3-319-73132-2_10

class. We clearly need to write methods to both access and set values of the array
We shall insist that the size of the array is set through a constructor. As such, we
shall not allow the user to change this variable through any method, but will write a
public method that allows us to access the size of a given vector.

Listing 10.1 `Vector.hpp`

```cpp
#ifndef VECTORHEADERDEF
#define VECTORHEADERDEF

class Vector
{
private:
  double* mData; // data stored in vector
  int mSize; // size of vector
public:
  Vector(const Vector& otherVector);
  Vector(int size);
  ~Vector();
  int GetSize() const;
  double& operator[](int i); // zero-based indexing
  // read-only zero-based indexing
  double Read(int i) const;
  double& operator()(int i); // one-based indexing
  // assignment
  Vector& operator=(const Vector& otherVector);
  Vector operator+() const; // unary +
  Vector operator-() const; // unary -
  Vector operator+(const Vector& v1) const; // binary +
  Vector operator-(const Vector& v1) const; // binary -
  // scalar multiplication
  Vector operator*(double a) const;
  // p-norm method
  double CalculateNorm(int p=2) const;
  // declare length function as a friend
  friend int length(const Vector& v);
};

// Prototype signature of length() friend function
int length(const Vector& v);

#endif
```

Listing 10.2 `Vector.cpp`

```cpp
#include <cmath>
#include <iostream>
#include <cassert>
#include "Vector.hpp"

// Overridden copy constructor
// Allocates memory for new vector, and copies
```

```
 8  // entries of other vector into it
 9  Vector::Vector(const Vector& otherVector)
10  {
11    mSize = otherVector.GetSize();
12    mData = new double [mSize];
13    for (int i=0; i<mSize; i++)
14    {
15      mData[i] = otherVector.mData[i];
16    }
17  }
18
19  // Constructor for vector of a given size
20  // Allocates memory, and initialises entries
21  // to zero
22  Vector::Vector(int size)
23  {
24    assert(size > 0);
25    mSize = size;
26    mData = new double [mSize];
27    for (int i=0; i<mSize; i++)
28    {
29      mData[i] = 0.0;
30    }
31  }
32
33  // Overridden destructor to correctly free memory
34  Vector::~Vector() {
35    delete[] mData;
36  }
37
38  // Method to get the size of a vector
39  int Vector::GetSize() const
40  {
41    return mSize;
42  }
43
44  // Overloading square brackets
45  // Note that this uses 'zero-based' indexing,
46  // and a check on the validity of the index
47  double& Vector::operator[](int i)
48  {
49    assert(i > -1);
50    assert(i < mSize);
51    return mData[i];
52  }
53
54  // Read-only variant of []
55  // Note that this uses 'zero-based' indexing,
56  // and a check on the validity of the index
57  double Vector::Read(int i) const
58  {
```

```
59      assert(i > -1);
60      assert(i < mSize);
61      return mData[i];
62    }
63
64    // Overloading round brackets
65    // Note that this uses 'one-based' indexing,
66    // and a check on the validity of the index
67    double& Vector::operator()(int i)
68    {
69      assert(i > 0);
70      assert(i < mSize+1);
71      return mData[i-1];
72    }
73
74    // Overloading the assignment operator
75    Vector& Vector::operator=(const Vector& otherVector)
76    {
77      assert(mSize == otherVector.mSize);
78      for (int i=0; i<mSize; i++)
79      {
80        mData[i] = otherVector.mData[i];
81      }
82      return *this;
83    }
84
85    // Overloading the unary + operator
86    Vector Vector::operator+() const
87    {
88      Vector v(mSize);
89      for (int i=0; i<mSize; i++)
90      {
91        v[i] = mData[i];
92      }
93      return v;
94    }
95
96    // Overloading the unary - operator
97    Vector Vector::operator-() const
98    {
99      Vector v(mSize);
100     for (int i=0; i<mSize; i++)
101     {
102       v[i] = -mData[i];
103     }
104     return v;
105   }
106
107   // Overloading the binary + operator
108   Vector Vector::operator+(const Vector& v1) const
109   {
```

```
110      assert(mSize == v1.mSize);
111      Vector v(mSize);
112      for (int i=0; i<mSize; i++)
113      {
114         v[i] = mData[i] + v1.mData[i];
115      }
116      return v;
117   }
118
119   // Overloading the binary - operator
120   Vector Vector::operator-(const Vector& v1) const
121   {
122      assert(mSize == v1.mSize);
123      Vector v(mSize);
124      for (int i=0; i<mSize; i++)
125      {
126         v[i] = mData[i] - v1.mData[i];
127      }
128      return v;
129   }
130
131   // Overloading scalar multiplication
132   Vector Vector::operator*(double a) const
133   {
134      Vector v(mSize);
135      for (int i=0; i<mSize; i++)
136      {
137         v[i] = a*mData[i];
138      }
139      return v;
140   }
141
142   // Method to calculate norm (with default value p=2)
143   // corresponding to the Euclidean norm
144   double Vector::CalculateNorm(int p) const
145   {
146      double norm_val, sum = 0.0;
147      for (int i=0; i<mSize; i++)
148      {
149         sum += pow(fabs(mData[i]), p);
150      }
151      norm_val = pow(sum, 1.0/((double)(p)));
152      return norm_val;
153   }
154
155   // MATLAB style friend to get the size of a vector
156   int length(const Vector& v)
157   {
158      return v.mSize;
159   }
```

The files required for the vector class are given above. These files may be down-loaded from https://www.springer.com/9783319731315. Subsequent sections of this chapter provide a commentary on why we have chosen to write the methods in the way in which they appear.

10.2 Constructors and Destructors

In the tip given in Sect. 4.3.3, we encouraged the reader to ensure that, when dynami-cally allocating memory, every new statement was matched by a delete statement. We explained that if this is not done, then the code may consume large amounts of the available memory. Eventually the computer will run out of memory, preventing the code (and any other application running) from proceeding any further. We have repeated this tip on several occasions. Writing appropriate constructors and destruc-tors for the vector and matrix classes allows us to automatically match a delete statement (through the calling of a destructor when the object goes out of scope) with every new statement (hidden from the user of the class in a constructor). We now describe appropriate constructors and a destructor for the class of vectors.

10.2.1 The Default Constructor

We want a constructor for the Vector class to allocate the memory required to store a given vector when it is called. The default constructor takes no arguments, and therefore this constructor has no way of knowing how many entries the vector requires. As such, it cannot allocate an appropriate size to the vector, and so we ensure that a default constructor is never used by not supplying a default constructor. The automatically generated default constructor will not be available to the user because we are supplying an alternative specialised constructor.

10.2.2 The Copy Constructor

Let us suppose we have an instance of the class Vector called u. If we were to use the automatically generated copy constructor to create another vector called v, then this constructor would *not* perform the tasks that we require of the copy constructor. The member mSize would be correctly set. However, the automatically generated copy constructor would not allocate any memory for the new copy of the data, and so it would be impossible for the entries of the vector to be copied correctly. What would actually happen is that the pointer mData in the original vector u would be assigned to the pointer mData in the new vector v. As no new memory would be allocated, this would have the effect that v would simply become a different name for the original vector u: there would only be one vector stored, and changing the entries of v would therefore have the unintended effect of changing those of u, and vice versa. A further complication of not overriding the default copy constructor would be that, because two vectors alias their mData pointers with the same piece of memory, both vectors would attempt to de-allocate it (by calling delete in their destructor, see Sect. 10.2.4) when they went out of scope.

What we actually want to happen when the copy constructor is called is for the member mSize of the new vector v to be set to the same value as for the original vector u. Memory should then be allocated for the new vector so that v has the same number of entries as u, and the entries of u then copied into the correct position in the new vector v. We therefore override the automatically generated copy constructor so that it sets the size of v to the size of u, allocates memory for the vector v of the correct size, and then copies the entries of u into v.

10.2.3 A Specialised Constructor

We have supplied no definition for the default constructor to ensure that it is never used, and have overridden the copy constructor so that if we already have a vector we may create a copy of that vector. We also include a constructor that requires a positive integer input that represents the size of the vector. This constructor sets the member mSize to this value, allocates memory for the vector, and initialises all entries to zero.

10.2.4 Destructor

The automatically generated destructor will delete the pointer mData and the integer mSize when an instance of the class Vector goes out of scope, but will not free the memory allocated to this instance of the class: this would be similar to not providing a matching delete statement for a new statement. We therefore override the automatically generated destructor to free the memory allocated for an instance of the class Vector when it goes out of scope.

10.3 Accessing Private Class Members

In Sect. 10.1 we explained that we were going to make both the size of the vector, mSize, and the pointer to the entries of the vector, mData, private members of the class. This has the advantage that we can only set the size of the vector through the constructor (ensuring that this member is a positive integer, and preventing us from inadvertently changing it while a code is being executed), and allows us to perform a validation that the index of an entry of a vector is correct before attempting to access that entry. In this section, we explain how we have written the methods that allow us to access these private members.

10.3.1   Accessing the Size of a Vector

The size, or length, of a vector is accessed through the public method `GetSize`
This member takes no arguments, and returns the private member `mSize`.

10.3.2   Overloading the Square Bracket Operator

We overload the square bracket operator so that, if `v` is a vector, then `v[i]` returns
the entry of `v` with index `i` using zero-based indexing. This method first checks that
the index falls within the correct range—that is, a nonnegative integer that is less
than `mSize`—and then returns a reference to the value stored in this entry of the
vector.

10.3.3   Read-Only Access to Vector Entries

The overloaded square bracket operator can be used for both reading data from the
vector and for changing entries of the vector, through a reference. Since we may
need to guarantee that some functions which read from a vector do not change it, we
also supply a read-only `const` version. This public method `Read` is similar to the
square bracket operator. It uses zero-based indexing and first checks that the index
falls within the correct range and then returns a copy of the value stored in this entry
of the vector.

10.3.4   Overloading the Round Bracket Operator

The round bracket operator is overloaded to allow us to access entries of a vector using
one-based indexing. We have chosen the round bracket operator for this purpose as
this allows similar notation to that employed by Fortran and MATLAB, both of which
use one-based indexing. In common with the overloaded square bracket operator
described in Sect. 10.3.2, this method first validates the index before returning the
appropriate entry of the vector.

10.4   Operator Overloading for Vector Operations

Readers with experience of programming in MATLAB will appreciate the feature of
this system that allows the user to write statements such as "`v = -w;`" and "`a =
b + c;`" where `v`, `w`, `a`, `b`, `c` are vectors of a suitable size. We will allow similar
looking code to be written for the vectors developed in this chapter through operator
overloading: i.e. we will define the assignment operator, and various unary and binary
operators. This will be very similar to the operator overloading for complex numbers

in Sect. 6.4. An additional feature required for the class being written here is a check that the vectors are all of the correct size: this will be enforced using `assert` statements.

10.4.1 The Assignment Operator

The overloaded assignment operator first checks that the vector on the left-hand side of the assignment statement is of the same size as the vector on the right-hand side. If this condition is met, the entries of the vector on the right-hand side are copied into the vector on the left-hand side.

10.4.2 Unary Operators

The overloaded unary addition and subtraction operators first declare a vector of the same size as the vector that the unary operator is applied to. The entries of the new vector are then set to the appropriate value before this vector is returned. Note that in the example statement "`v = -w;`" above, it is the assignment operator's responsibility to check that sizes of v and w match and the unary subtraction need do no error checking.

10.4.3 Binary Operators

The overloaded binary operators first check that the two vectors that are operated on are of the same size. If they are, a new vector of the same size is created. The entries of this new vector are assigned, and this new vector is then returned. In the example statement "`a = b + c;`" above, it is the binary addition operator's responsibility to check that the sizes of the vectors b and c match, but the assignment operator's responsibility to check that the result can safely be assigned to a.

10.5 Functions

A function to calculate the p-norm of a vector is included in our class of vectors. See Sect. A.1.5 for a definition of the p-norm of a vector. This implementation allows the user to call the function with an optional argument p: if this is not specified the default value $p = 2$ (corresponding to the Euclidean norm) will be used.

10.5.1 Members Versus Friends

We note that most functionality in the class is given via member methods and member operators. In order to calculate the 2-norm of a vector or to inspect its size, we must write "u.CalculateNorm();" or "u.GetSize();", respectively. This may be considered a clumsy syntax by some users, especially those with experience of MATLAB, and so we provide an alternative length function to complement the GetSize method. The length function is declared as a *friend* within the class which enables it to read the private mSize member. Note that whereas many of the members of the class are declared const at the end of the signature—to ensure they do not change the class itself—the length function guarantees that the vector which it is given as an argument will remain constant through making the argument a constant reference variable.

10.6 Tips: Memory Debugging Tools

We stressed in a previous tip (Sect. 4.3.3) that every new should be matched with a delete. This is especially important when a program allocates memory within a loop. If a long-running program repeatedly allocates memory without de-allocating it, then eventually that program will unnecessarily occupy all the available memory of the computer. This problem—known as a *memory leak*—will eventually cause the program to fail.

There are memory-related problems other than memory leakage. The following code illustrates some common memory errors. The loop in lines 8–11 has an incorrect upper bound and thus the program attempts to write to x[10] which does not match the 10 elements allocated to x in line 3. The variable z is never initialised, which means that the flow of the program at the if statement on line 15 is unpredictable. The second delete statement—on line 23—is in error since it attempts to de-allocate memory which has already been de-allocated on the previous line. Finally the memory for y which was allocated on line 4 is never deleted.

Listing 10.3 Broken.cpp

```
1   int main(int argc, char* argv[])
2   {
3     double* x = new double[10];
4     double* y = new double[10];
5
6     // Error: x[10] is accessed
7     // May cause a run-time error
8     for (int i=0; i<=10; i++)
9     {
10       x[i] = i;
11     }
12
13     // Error: z is not set
14     int z;
15     if (z == 0)
16     {
17       y[0] = x[0];
18     }
19
```

```
20      // Error: x de-allocated twice
21      // May cause a run-time error
22      delete[] x;
23      delete[] x;
24      // Error: y still allocated
25   }
```

The four problems in the program above will not prevent the code from being compiled. The program may also run as expected until the final `delete` statement, but crash at that point. So, in this program, most of the memory errors are undetectable in normal circumstances.

These errors can be detected with a memory debugging tool such as the open source programs *Valgrind* or *Electric Fence*. These tools run an executable file while inspecting all the memory access calls. Some tools (such as Electric Fence) do this by replacing the usual memory libraries with ones which intercept the calls. Others tools (such as Valgrind) run the program inside a virtual machine and externally monitor the memory accesses—a slower process, but one which does not require recompilation of the program.

On running the program given in Listing 10.3 through Valgrind all four memory problems are detected. A summary of the Valgrind output is given below.

```
Invalid write of size 8
   at 0x4006BA: main (Broken.cpp:10)

Conditional jump or move depends on uninitialised value(s)
   at 0x4006D1: main (Broken.cpp:15)

Invalid free() / delete / delete[]
   by 0x4006F8: main (Broken.cpp:23)

80 bytes in 1 blocks are definitely lost...
   by 0x40069A: main (Broken.cpp:4)
```

10.7 Exercises

The exercises in this chapter guide you to build on the `Vector` class with an additional `Matrix` class. These classes are then combined into a `LinearSystem` class (or, in the final exercise, an alternative class derived from it) which has a method for solving systems of the form $\mathbf{Ax} = \mathbf{b}$ for $\mathbf{x}$. Example solutions for these classes are given in Sect. C.1. Figure 10.1 illustrates a typical solution to these exercises with a collaboration diagram for all the classes produced by these exercises. This diagram uses the same UML syntax as Fig. 7.1, as described in Sect. 7.2.

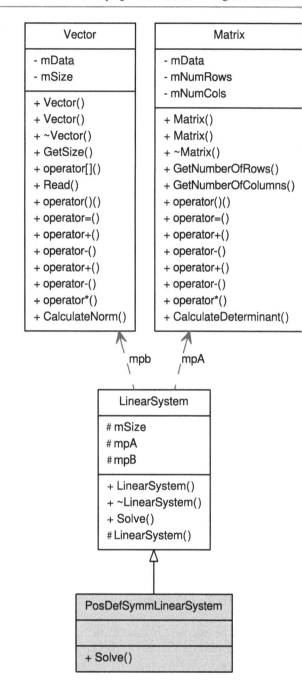

Fig. 10.1 Class collaboration diagram for PosDefSymmLinearSystem

The files Vector.hpp and Vector.cpp given in Listings 10.1 and 10.2, as well as the example Matrix and LinearSystem classes given in Sect. C.1, may be downloaded from https://www.springer.com/9783319731315.

10.1 Write a suitable suite of tests to black box test the class of vectors.

10.2 Make any improvements you might deem appropriate to the class of vectors. You might be helped in this task by the following list.
- The assertions for the round bracket operator are almost identical to those of the square bracket operator and those of the Read method. Rewrite the Read method and one of these operators in such a way that they call the remaining operator (with a suitable offset, as necessary) and all the checks are given in one place.
- There are many assertions in the class as it stands. These mean that it is very easy to write programs which terminate with a run–time error. Can you turn any of the assertions into exceptions or warnings (see Chap. 9)?
- Write an output operator for vectors using the pattern given in Sect. 6.4 for the operator<< in the complex number class.

10.3 In this exercise, we will develop a class of matrices called Matrix for use with the class of vectors developed in this chapter. The class of matrices should include the features listed below. Your class should have private members mNumRows and mNumCols that are integers and store the number of rows and columns, and mData that is a pointer to a pointer to a double precision floating point variable, which stores the address of the pointer to the first entry of the first row. See Appendix A for details of the linear algebra that underpins these operations. Use a suitable testing strategy when developing this class.

1. An overridden copy constructor that copies the variables mNumRows and mNum-Cols, allocates memory for a new matrix, and copies the entries of the original matrix into the new matrix.
2. A constructor that accepts two positive integers—numRows and numCols—as input, assigns these values to the class members mNumRows and mNumCols, allocates memory for a matrix of size mNumRows by mNumCols, and initialises all entries to zero.
3. An overridden destructor that frees the memory that has been allocated to the matrix.
4. Public methods for accessing the number of rows, and the number of columns.
5. An overloaded round bracket operator with one-based indexing for accessing the entries of the matrix so that, provided i and j are valid indices for the matrix, A(i,j) may be used to access mData[i-1][j-1].
6. Overloaded assignment, unary and binary operators to allow addition, subtraction and multiplication of suitably sized matrices, vectors and scalars. You should use assert statements to ensure the matrices and vectors are of the correct size.
7. A public method that computes the determinant of a given square matrix.

10.4 In this exercise, we will develop a class called LinearSystem that may be used to solve linear systems. Assuming the system is nonsingular, a linear system is defined by the size of the linear system, a square matrix, and vector (representing the right-hand side), with the matrix and vector being of compatible sizes. The data associated with this class may be specified through an integer variable mSize, a pointer to a matrix mpA, and a pointer to the vector on the right-hand side of the linear system mpb. We suggest only allowing the user to set up a linear system through the use of a constructor that requires specification of the matrix and vector: the member mSize may then be determined from these two members. If you do not wish to provide a copy constructor, then the automatically generated copy constructor should be overridden and made private to prevent its use. As with the class of vectors, we recommend that use of the automatically generated default constructor is prevented by providing a specialised constructor but no default constructor. A public method Solve should be written to solve this linear system by Gaussian elimination with pivoting, as described in Sect. A.2.1.3. This method should return a vector that contains the solution of the linear system.

Test your class using suitable examples. We suggest that you write a set automated tests in a testing framework such as CxxTest. An outline model test-suite for testing the linear algebra classes is given in Listing C.5 in Section C.1. When considering what to test think about the following.

- How you might test the various constructors.
- How you will black box test solving problems when the matrix is poorly conditioned.
- How to test that the Gaussian Elimination routine is performing pivoting when small values appear on the diagonal.
- How to test various Matrix and Vector methods.

10.5 Derive a class called PosDefSymmLinearSystem (or similar) from the class LinearSystem that may be used for the solution of positive definite symmetric linear systems. Make the method Solve a virtual method of the class Linear-System, and override this method in the class PosDefSymmLinearSystem so that it uses the conjugate gradient method for solving linear systems described in Sect. A.2.3. If you declared LinearSystem member data as private in the previous exercises, then this should now be declared protected. Your class PosDefSymmLinearSystem should perform a check that the matrix used is symmetric: testing that the matrix is positive definite would be rather difficult and so we don't suggest performing a check for this property. Test your class using suitable examples.

An Introduction to Parallel Programming Using MPI

11

This chapter serves as an introduction to the *Message Passing Interface* (MPI), which is a widely used library of code for writing parallel programs on distributed memory architectures. It is not intended that you will learn much about parallel programming from reading this chapter—we would recommend that you use a dedicated textbook (such as those listed in the Further Reading section at the end of this book [9, 10]) or tutorial if you wish to gain a detailed knowledge. However, this chapter should give you a *basic guide* to compiling and running parallel programs written using MPI. If you are likely to use a scientific library built on MPI (such as PETSc[1]) then what you learn here in this chapter should help to demystify some of the library calls, and enable you to begin to edit parallel programs written by other programmers.

11.1 Distributed Memory Architectures

There are several ways of classifying parallel computers and parallel programs but the most basic one is that of *shared memory* versus *distributed memory* machines. In the shared memory architecture several processing units (often called "cores" nowadays), share access to a common pool of memory, as shown in Fig. 11.1. This architecture has the advantage that a part of a program running on one core can easily communicate with another, since it can read or write in the memory space of the other part of the program. Programming for shared memory can be quite easy and the programs are generally very quick, but historically shared memory machines

---

[1]The Portable Extensible Toolkit for Scientific Computing (PETSc, pronounced "pet see") is a library providing functionality for the solution of linear and nonlinear systems of equations on both sequential and parallel architectures.

© Springer International Publishing AG, part of Springer Nature 2017

J. Pitt-Francis and J. Whiteley, *Guide to Scientific Computing*
in C++, Undergraduate Topics in Computer Science,
https://doi.org/10.1007/978-3-319-73132-2_11

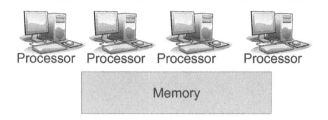

Fig. 11.1 A shared memory parallel architecture: the processors/cores are co-located and share a common memory

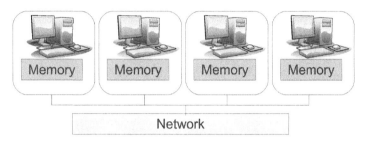

Fig. 11.2 A distributed memory parallel architecture: each processor has sole access to its local memory and the machines are connected on the same network

have been expensive, require specialised hardware, and physical constraints limit the total number of cores. This situation is, however, now changing as most new desktop computers have two or more cores.

The other main architecture commonly used for parallel programming is the distributed memory architecture (see Fig. 11.2), where each processing unit has a local memory space where it can read and write with ease, but the memory of other processors is completely hidden. The processors are connected—allowing data to be communicated between processors—on a network which could be a dedicated fast switch network (in the case of a cluster computer) or could be the wider Internet. The existence of the network between the processing units means that programming for this architecture is likely to be more complicated, and that programs that rely heavily on communication between processors are likely to be slower. However, as we shall now explain, distributed memory programs are versatile.

The versatility of distributed memory programs is evidenced by the fact that it is possible to take a program intended for a distributed memory architecture and run it on a shared memory architecture. In this case, the individual parts of the parallel program will have separate memory spaces within the shared memory system (so that they cannot directly access each others' memory), but will be able to communicate via the memory system. Communication is therefore much faster than over a network. Distributed memory programs can readily be executed on shared memory machines and are fast, but are also memory-hungry. The reverse is not true: you cannot, in general, run a shared memory program on a distributed memory cluster.[2]

You can even run distributed memory programs on a computer with a single processor. All the parallel processes will be run as what are known as individual *threads*, and will communicate via the memory system with the operating system responsible for context switching between the threads. There is no performance advantage to doing this, since there is an overhead to run many threads on a single processor. The advantage is that you can write and debug a program on a low-powered laptop, tune it on a shared memory desktop, and then deploy exactly the same program on a supercomputer.

11.2 Installing MPI

MPI is actually a set of standards for performing distributed computing. The MPI-1 standard documents the primary core of MPI (basic point-to-point and collective communication) while the MPI-2 standard adds other useful but advanced features such as parallel file access (through the input and output operations provided by MPIIO) and remote memory access (one-sided communication). Because MPI is a set of open standards there are various implementations available to choose from. The most commonly used are the MPICH and Open MPI implementations. The current versions of MPICH and Open MPI (formerly LAM/MPI) both implement all the functionality in both the MPI-1 and MPI-2 standards.

Both MPICH and Open MPI are open source projects, under active development and freely available to download. They may be run on a wide variety of machine architectures, operating systems and communication infrastructures. The Open MPI library implementation is currently available from major Linux distribution repositories and is therefore easy to install on Linux systems. It is configured so that it can be used either on a stand-alone system (in the manner of a shared-memory system) or across standard Ethernet using the secure shell `ssh` protocol.

11.3 A First Program Using MPI

Just as in Sect. 1.2, we introduce the MPI library by using a program that prints the text "Hello World" to the screen. This time, it runs and prints in parallel. This simple example C++ MPI program is shown below.

---

[2]There are several programming libraries which allow the programmer access to a *distributed shared memory* computer where machines over a network act as if they were part on one contiguous system. There has, however, not been wide-spread use of these libraries at the time of writing.

Before explaining the purpose of the individual statements in this program, we need to explain what we mean by the term *process*. Loosely speaking a process is a part of a parallel program that may be executed independently of the other parts provided that data can be communicated through MPI calls when required. As such a process can be thought of as a component of the program that can be executed on one of the processors shown in Fig. 11.2. (However, we make a distinction between processes and physical processors—or cores—because it is possible to run multiple processes on a single processor.) If a code has p processes, then each process is given a rank which is a unique integer in the range $0 \le$ rank $< p$.

Listing 11.1 MpiHelloWorld.cpp

```
1   #include <iostream>
2   #include <mpi.h>
3
4   int main(int argc, char* argv[])
5   {
6       MPI::Init(argc, argv);
7
8       int num_procs = MPI::COMM_WORLD.Get_size();
9       int rank = MPI::COMM_WORLD.Get_rank();
10      std::cout << "Hello world from process " << rank
11               << " of " << num_procs << "\n";
12      MPI::Finalize();
13      return 0;
14  }
```

There are several lines of the program above which mention MPI. The first of these is the extra include on line 2 which allows the program to see the full functionality of the MPI library. Subsequently, there are MPI::Init and MPI::Finalize statements on lines 6 and 12 which start and stop the parallel part of the code. All MPI calls must lie between these two statements. The method Get_size allows us to access the number of processes taking part in the program execution, and the method Get_rank allows us to identify the process which is executing a given statement. The COMM_WORLD object represents a communications group involving *all the processes* running the current calculation. It is possible to split this communication group up into smaller groups so that subsets of the processes can share private data.

It should be noted that all the calls to MPI in program above use calls to specific C++ bindings to the MPI library. So Finalize is a function in the MPI *namespace* (see Sect. B.4) and Get_size is a method of the communication object COMM_WORLD. Some C++ programmers prefer not to use these bindings, but opt instead for the plain C functions MPI_Init, MPI_Get_size, etc. which have a slightly different syntax. Both versions are valid in C++ programs and can even be mixed.

11.3.1 Essential MPI Functions

The functions `MPI::Init` and `MPI::Finalize` on lines 6 and 12 of Listing 11.1 are required calls in any MPI program. On line 6, `MPI::Init` is able to promote the program from a single executable to a parallel program running as several processes. In order to do this, it needs to know how many processes to launch and on which machines they should be run—as we shall see in Sect. 11.3.2 this is information that can be made available via command-line arguments. `MPI::Init` inspects the command-line arguments provided by `argc` and `argv`, acting on any it recognises.

Some MPI implementations of `MPI::Init` update their arguments by removing those which have been acted upon. Therefore, if you want an MPI program to read some specific arguments from the command-line for use in your calculation, the best place to do this is *after* `MPI::Init` since, at that point, all MPI-specific arguments have been read and updated as necessary.

The `MPI::Finalize` makes sure that the program closes down neatly, closing any remote connections and terminating all processes.

11.3.2 Compiling and Running MPI Code

So far when we have compiled C++ programs we have included code either from standard locations (through including files such as `cmath`) or from other parts of our own code (such as `Book.hpp`). MPI, as a *third-party library*, is not part of the standard C++ distribution.

Normally when you compile against a third-party library, you would have to include extra compiler flags specifying the location of the header files, the location of the libraries themselves, and names of some of the library dependencies. This can be a little onerous. Added to this, on some large computing facilities there may be several versions of MPI available which make it possible to accidentally compile some of your program with one version, and the remainder of the program with another, possibly incompatible, version. To ameliorate these difficulties, the MPI distributions have provided "wrapper compilers" for C++ (as well as for C and Fortran). The wrapper compiler automatically adds the correct compiler flags when it calls the actual compiler. The C++ MPI compiler on most systems will be called `mpiCC`, `mpic++` or `mpicxx`. It is probably the case that it exists with more than one synonym.

The standard Linux distribution of the Open MPI package has an `mpiCC` compiler which is a wrapper to the GNU `g++` C++ compiler. To ensure that this compiler is installed, open a terminal window and type "`which mpiCC`" followed by return. Hopefully the computer will respond by reporting the location of this compiler, for example,

```
$ which mpiCC
/usr/bin/mpiCC
$
```

To compile the code given in Listing 11.1, open a terminal window and create a directory where code may be saved. Move into this directory, and save the code as "MpiHelloWorld.cpp". The MPI wrapper compiler may have some compiler flags of its own, but most flags are passed on to the normal g++ compiler. In the same directory type,

```
mpiCC -o MpiHelloWorld MpiHelloWorld.cpp
```

It is possible (but uninteresting) to run the executable which you have just produced as a standalone program. That is, without any of the MPI machinery and with no code run in parallel. If this is the case then Get_size will return 1, and, because there is only one process, Get_rank on that process will return 0. Just as in array numbering, the process rank numbering starts at zero, so each process is given a unique integer rank in the range $0 \leq$ rank $< p$, where p is the total number of processes.

```
$ ./MpiHelloWorld
Hello world from process 0 of 1
```

To run in parallel either on the same machine or across a network or cluster, use the mpirun command (also known as mpiexec on some MPI implementations). This command will, if necessary, launch a service (called a daemon) on all the machines involved in the calculation. It will then make sure that copies of the executable can be run on every machine.

To run the program locally, use the "number of processes" -np flag.

```
$ mpirun -np 2 ./MpiHelloWorld
Hello world from process 0 of 2
Hello world from process 1 of 2
```

To run the program across a network you can give a list of machines in a *host file*, or alternatively list the machine names on the command-line. It is imperative that you have an account on the remote machines, that you are able to connect via ssh (preferably without being prompted for a password), that the machines have the same MPI implementation installed on them, and that they are capable of running the executable file which you are sending. In the example below, ranks 0 and 2 of a 3-process job are launched on remote machines. The rank 1 process will run on the local machine, from where the job has been launched. Note that in this case buffered output from the local machine has appeared on the screen before output which has been sent from the remote machines.[3]

```
$ mpirun -host remote1.org,localhost,remote2.org ./MpiHelloWorld
Hello world from process 1 of 3
Hello world from process 2 of 3
Hello world from process 0 of 3
```

If you are running your program on a large cluster or a supercomputer, then it is likely that the program will be launched from a script via a queueing system. In this case, the locations of the processors available to you will be determined by the job queue manager. You should obtain detailed instructions from the system administrators about which arguments to give to the mpirun command in your script.

11.4   Basic MPI Communication

While the parallel "Hello World" program used the MPI libraries, it did not make any use of the communication features offered by these libraries. More specifically, it did not do any *message passing*, which is the main feature of MPI. In this section, we give a brief survey of some of the common communication patterns available in the MPI library through providing a sample of the large range of available function calls.

11.4.1   Point-to-Point Communication

The essential part of MPI functionality is being able to send a single message between processes, where one process *sends* while another process *receives*. These two functions are called Send and Recv. Their function prototypes are:

```
void Comm::Send(const void* buf, int count,
                const Datatype& datatype,
                int dest, int tag) const
```

---

[3]MPI implementations vary in how they return console input from the individual processes to the console from which the program was launched. Even when flush is called on the cout stream it may still be the case that the MPI machinery is buffering output.

```
void Comm::Recv(void* buf, int count,
                const Datatype& datatype,
                int source, int tag) const
```

The Send method takes data on the current process from the location given by the pointer buf. These data are assumed to be in contiguous memory (as an array of count variables), but buf may be a pointer to a single variable. Note the const keyword next to the buffer argument: MPI is making a guarantee not to alter your data during the message sending. The datatype field tells the system what the type of the data is (so that the correct number of bytes are sent in the correct format). The last two arguments of the Send method give the destination process number (this is the *rank* of the process we wish to send to) and a *tag*. The message tag can be any nonnegative integer value and its purpose is to allow the user to easily identify the context of a message. Negative tag values are reserved by the library for special values such as MPI::ANY_TAG which is introduced below.

The Recv method has the same basic arguments: a pointer to a buffer in which to store the message, an integer count that gives the expected number of items in the message, the data-type for these items, the rank of the source process which is sending the message and the tag value of a message. The receiver is allowed to use wild-cards for either the source of the message, or the message tag, or both. The wild-card MPI::ANY_SOURCE[4] is useful if, for example, we wish to receive all the results of one phase of computation (tagged with phase_1_tag, for example) before moving on to the next phase. Messages sent with the next tag (phase_2_tag) can then be queued until the receiving process is ready for them. The wild-card MPI::ANY_TAG is useful if we know which process is sending the data, but do not know what the tag will be.

The corresponding MPI Datatype signatures for the types introduced in Chap. 1 are MPI::BOOL, MPI::CHAR, MPI::INT and MPI::DOUBLE.[5] There is no MPI type for strings because std::string is a C++ class rather than a plain data-type. It is possible to send entire C++ classes in MPI messages by using advanced programming features to introduce user-defined data-types, but this is not recommended. Classes can readily be transferred by packing the raw data into a message at one end and unpacking it into a waiting class at the other end.

The following code fragment illustrates sending one message consisting of two floating-point numbers from process 0 to process 1. Note that code involving point-to-point communication is necessarily nonsymmetric: both processes are running exactly the same program with the same code, but parts of the program which are intended only for one process are placed in specific blocks guarded by their process rank.

---

[4]MPI::ANY_TAG and MPI::ANY_SOURCE are C++ names for these wild-card values. Many codes use the interchangeable C names: MPI_ANY_TAG and MPI_ANY_SOURCE.

[5]Note that these are the C++ object names for these types—they are also called synonymously by their C names: MPI_BOOL, MPI_CHAR, MPI_INT and MPI_DOUBLE.

Listing 11.2 Example code for sending and receiving using the MPI libraries

```
1   int tag = 30;
2   if (MPI::COMM_WORLD.Get_rank() == 0)
3   {
4       //Specific send code for process 0
5       double send_buffer[2] = {100.0, 200.0};
6       MPI::COMM_WORLD.Send(send_buffer, 2,
7                               MPI::DOUBLE, 1, tag);
8   }
9   if (MPI::COMM_WORLD.Get_rank() == 1)
10  {
11      //Specific receive code for process 1
12      double recv_buffer[2] = {0.0, 0.0};
13      MPI::COMM_WORLD.Recv(recv_buffer, 2, MPI::DOUBLE,
14                              MPI::ANY_SOURCE, MPI::ANY_TAG);
15      std::cout << recv_buffer[0] << "\n";
16      std::cout << recv_buffer[1] << "\n";
17  }
```

11.4.1.1   Blocking and Buffered Sends

The default means of sending point-to-point messages with `Send` and `Recv` represents one combination in a spectrum of available communication protocols. Both functions are known as *blocking* functions because they do not allow the execution of the program to continue until it is safe to do so. The `Send` method not only guarantees that it will not change the contents of the data buffer, but that any subsequent changes to the data buffer will not affect the message that is being sent. So if computation is allowed to proceed from a `Send` call it either means that the message has already been delivered or that the data has been copied into another buffer ready for delivery.

The default `Send` is a compromise between the safety of waiting to be sure that a message has been delivered and the efficiency of getting on with other tasks after sending the message immediately. The other send functions have similar function prototypes, but slightly different names. We briefly describe these send functions below: the interested reader should consult a dedicated MPI programming book (such as [9, 10]) for more details.

- The very safest, but possibly most inefficient, means of sending a message is to use a *blocking* synchronous send, `Ssend`. This function guarantees not to continue until the message has been delivered. This is a little like delivering a message by telephone conversation, because we cannot get on with our lives until the call has been made and the information has been relayed.
- A slightly more configurable version is `Bsend`, the *buffered* send. Like the plain `Send`, it allows the program to continue when safe, but this may happen faster since the message is copied to a separate buffer. This buffer must be supplied and configured by the user.

- At the top end of the spectrum, the most efficient means of sending a message is the *immediate* send Isend, which returns control to the program immediately whether the message has been delivered, buffered or not yet acted on. This is a little like communicating via SMS text message in which we are able to press "send" and get on with other things safe in the knowledge that the recipient will get the information some time soon. Because it may be dangerous to overwrite the original data contained in the message, MPI provides functions for testing whether or not the message has been delivered. The Isend command gives back a handle (called an MPI::Request) which has a Wait method: this method instructs execution to "wait here" until the message has been sent.
- There are a few other flavours of Send including compatible combinations: an immediate send can make use of a user-supplied buffer using the buffered non-blocking combination Ibsend.

The default Recv function is also one of a spectrum of functions. It is technically a *blocking* function, because execution cannot continue until a suitable message has been received. There is also a *non blocking* immediate receive Irecv together with some utilities for probing whether there are any queued messages which match certain sources or tags. This means that your program, rather than waiting for messages to be received, could get on with useful work, occasionally going back to check for new information.

11.4.2 Collective Communication

Code for point-to-point communication is not symmetric: one process sends while another receives. MPI provides specialised collective calls in which all the processes take part by executing the same commands. There are several major different flavours of these communication patterns: the combined send-receive (where every process sends a message to another, while also receiving a remote message); one-to-many operations such as *broadcast* where data from one process are sent to the entire group; and many-to-one operations such as *reduction* in which an operation is used to combine results from all processes into a single result.

These collective calls have the advantage that they can be highly tuned in an MPI implementation to fit the local architecture. The broadcast of a single number to all p processes from process 0 could be achieved by sending $p - 1$ messages from process 0, one message to each of the recipients. However, if process 0 sends to only two processes who each send to two more, then the information is broadcast to all recipients in about $\log_2 p$ rounds of message sending. If a supercomputer consists of several multicore computers connected by Ethernet, then the broadcast algorithm can be tuned to minimise the number of Ethernet messages while possibly increasing the number of faster messages between cores in the same machine.

11.4.2.1 Barrier

The simplest collective method is `Barrier`. It says that every process should wait here until all processes are ready to proceed. Barriers are useful when you are timing certain parts of the code, printing out information to the console, or debugging the code. In the following code example there is a `Barrier` at line 3, the purpose of which is to ensure that all processes have completed writing their output (from line 1) before they are permitted to proceed.

```
1    std::cout << "Processes may arrive at any time\n";
2    std::cout.flush();
3    MPI::COMM_WORLD.Barrier();
4    std::cout << "All processes continue together\n";
5    std::cout.flush();
```

11.4.2.2 Combined Send and Receive

There are many cases when we might wish to send and receive many point-to-point messages at the same point in a computation, and where every process should be involved.

For example, consider solving a partial differential equation (PDE) using a finite difference scheme over a regular grid (such a grid is illustrated in two dimensions in Fig. 12.2) where the value of a variable at one position on the grid depends on the value of that variable at a few neighbouring grid points. A similar example from a different field is that of image processing: many image processing filters, such as edge detection or blurring, are implemented as weighted averages of image intensities over a small patch of neighbouring pixels. Such problems may be readily parallelised by dividing the grid or image into a number of identical vertical (or horizontal) strips and assigning one strip to each parallel process. Each process can compute its partition independently, except at the edges where information at grid-points or pixels assigned to the neighbouring process is required. A way to provide this information is to keep a local copy of the required neighbouring grid-point data and to update these data from the neighbouring process by message-passing. The local copy of remote neighbouring data is called *halo data* and the message passing process is called *halo exchange*.

Halo exchange is demonstrated in Fig. 11.3 using the example of an image processing filter. This filter produces an image where the image intensity in the processed image at a given pixel is the average of the image intensities in the original image at five pixels: the pixel of interest; the pixel to the left; the pixel to the right; the pixel directly above; and the pixel directly below. The pixels allocated to process n are those in the shaded area in Fig. 11.3. We may calculate the processed image at the pixels represented by open circles in this shaded region using only pixel intensities stored by process n. Before we may calculate the filtered image at the pixels represented by solid circles on the left edge of process n, however, we require access to the pixels represented by solid circles on the right edge of the pixels stored by

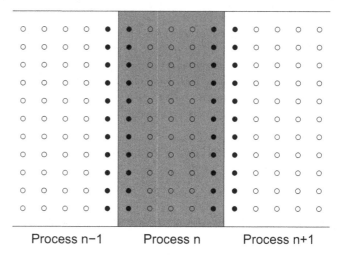

Fig. 11.3 Halo exchange between processes

process n-1: these pixels are referred to as the halo, and we need to copy these to process n before we can calculate the whole processed image. Similarly, the nodes along the left-hand boundary of process n must be copied to process n-1 before the processed image may be calculated. This procedure of sending edge data in both directions between processes n and n-1 is known as halo exchange. For the same reasons, two-way halo exchange is also required between processes n and n+1.

The partitioning of data between processes should ideally minimise the amount of data that has to be passed in halo exchanges: this is important when fine-tuning your code to produce optimum efficiency, but is beyond the scope of this book.

For these types of problem, a more sophisticated version of point-to-point message passing is the combined send and receive, called Sendrecv. Its function prototype is:

```
void Comm::Sendrecv(const void *sendbuf, int sendcount,
                    const Datatype& sendtype,
                    int dest, int sendtag,
                    void *recvbuf, int recvcount,
                    const Datatype& recvtype,
                    int source, int recvtag) const
```

Note that the ten arguments are divided into two sets of five: a set of send arguments about the outgoing message and a set of receive arguments about the incoming message. These are similar to the arguments given to the point-to-point versions in Sect. 11.4.1 and they are interpreted relative to the *local process*: if each process is sending to the rank above, by symmetry, each must be receiving from the rank below. It is possible to mix the types of messages (both in terms of DataType and

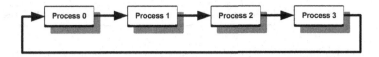

Fig. 11.4 Message passing between processes in a ring using combined send-receive

the length of the messages) so that, for example, odd-ranked processes are sending integer messages to the process above, but even-ranked processes are sending double precision floating point data. In this circumstance, on any given process the types of send and receive data will differ. As with the Recv functions, we can use the wild-cards for the source process identity and the received message tag.

The following code shows all processes communicating in a ring. Each process (with rank given by the variable rank) sends a message to its right-hand neighbour (rank + 1). Modular arithmetic—see Sect. 1.4.3—ensures that the left_rank and right_rank variables are set inside the range $0 \leq$ rank $<$ num_procs so that the top-most process is able to send a message to the rank 0 process. This message passing is illustrated schematically in Fig. 11.4 for four processes: the arrow indicates the direction in which the message is passed.

```
1   int tag = 30;
2   int rank = MPI::COMM_WORLD.Get_rank();
3   int num_procs = MPI::COMM_WORLD.Get_size();
4   // left_rank is rank-1
5   // Note modular arithmetic, so that 0 has
6   // neighbour num_procs-1
7   int left_rank = (rank-1+num_procs)%num_procs;
8   int right_rank = (rank+1)%num_procs;
9   int recv_data;
10  // Communicate in a ring ...->0->1->2...
11  MPI::COMM_WORLD.Sendrecv(&rank, 1, MPI::INT,
12                           right_rank, tag,
13                           &recv_data, 1, MPI::INT,
14                           left_rank, tag);
15  std::cout << "Process " << rank << " received from "
16            << recv_data << "\n";
```

There are cases, such as the halo exchange situation outlined above, where *nearly every* process will send halo data from the right-hand edge of its domain up to the next process to become a left-hand halo, but the top-most process does not need to send any data and the bottom-most process needs no left-edge halos. This is illustrated in Fig. 11.5 where four processes are taking part in the communication with the arrows indicating the direction in which information is passed. In a separate send-receive event the left-edges would also be sent down the chain to become right-edge halos, but again there is no need to send data from the bottom-most process. For this reason, MPI provides a special process name MPI::PROC_NULL which means that this process does not participate with a send and/or receive. This process name

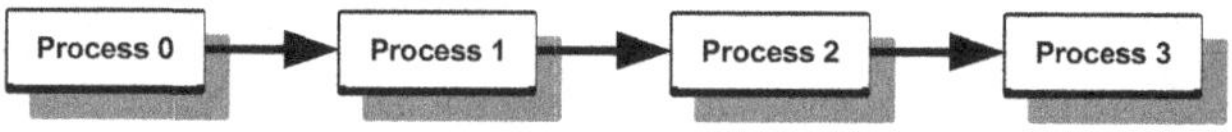

Fig. 11.5 Message passing between processes in a chain using combined send-receive. On process 3 the message destination is set to PROC_NULL

is illustrated in the following code, which is similar to the previous Sendrecv example, except that there is no closed loop: the top-most process does not send to process 0.

```
int tag = 30;
int rank = MPI::COMM_WORLD.Get_rank();
int num_procs = MPI::COMM_WORLD.Get_size();
int right_rank = rank+1;
// Top-most sends nowhere
if (rank == num_procs - 1)
{
    right_rank = MPI::PROC_NULL;
}
int left_rank = rank-1;
// Bottom-most receives nothing
if (rank == 0)
{
    left_rank = MPI::PROC_NULL;
}
int recv_data = 999; //This will be unchanged on proc 0
// Communicate 0->1->2... Final process sends nowhere
MPI::COMM_WORLD.Sendrecv(&rank, 1, MPI::INT,
                         right_rank, tag,
                         &recv_data, 1, MPI::INT,
                         left_rank, MPI::ANY_TAG);
std::cout << "Process " << rank << " received from "
          << recv_data << "\n";
```

11.4.2.3 Broadcast and Reduce

The collective operations broadcast and reduce are primarily one-to-many and many-to-one operations. In a broadcast (Bcast) operation, data from one process are shared with all other processes in the communication group. In a reduction operation, all the data is concentrated to a single process. This reduction operation is likely to be of one of a standard set available for numerical data (MPI::MAX, MPI::MIN, MPI::SUM, and MPI::PROD). There are other predefined reduction operations available including some bit-wise operations, and there is also opportunity to define extra operations. The prototype signatures of the broadcast and reduce operations are given below. Note that the argument root is the *source* of the broadcast but the *destination* of the reduction. MPI also provides a many-to-many reduction operation Allreduce which may be thought of as a reduction operation followed by a broadcast to all processes.

```
void Comm::Bcast(void* buffer, int count,
                 const MPI::Datatype& datatype,
                 int root) const

void Comm::Reduce(const void* sendbuf, void* recvbuf,
                  int count, const MPI::Datatype& datatype,
                  const MPI::Op& op, int root) const
```

An example reduction operation is given in Sect. 11.5.1 where the partial sums of a series are summed together in a single reduction step. For now, here is a broadcast example in which one process—process 0—mimics throwing three dice by generating integer random numbers from 1–6 inclusive, and broadcasts the results of all three throws. Each process then adds their own rank to the value shown on the first die, and a reduction operation reports on the maximum value attained after this operation.

```
1    int dice[3] = {0, 0, 0};
2    //Proc 0 sets the dice (#sides)
3    if (MPI::COMM_WORLD.Get_rank() == 0)
4    {
5        for (int i=0; i<3; i++)
6        {
7            dice[i] = (rand()%6)+1;
8        }
9    }
10   //Proc 0 broadcasts
11   MPI::COMM_WORLD.Bcast(dice, 3, MPI::INT, 0);
12   //Every process adds their rank to dice[0]
13   dice[0] += MPI::COMM_WORLD.Get_rank();
14   //Reduce the first value to get the maximum
15   int max;
16   MPI::COMM_WORLD.Reduce(dice, &max, 1,
17                              MPI::INT, MPI::MAX, 0);
18   //On Proc 0: max = dice[0]+MPI::COMM_WORLD.Get_size()-1
```

11.4.2.4 Scatter and Gather

The scatter and gather operations are extensions to broadcast and reduction operations. They are the most advanced operations which we cover in this book, and we do so because the gather operation is useful for taking data which has been distributed across processes and *concentrating* it onto a single process. For example, if a vector is split across processes in a similar manner to a PETSc vector we might wish to write it to a file using a single write operation using only one process.[6]

---

[6]There are a few standard ways of getting data to file from a parallel program: *concentration*, where one process does all the writing, as suggested above; *round-robin* where processes take it in turns to

The scatter operation Scatter is similar to the broadcast operation in that it is one-to-many with one process being responsible for sending the message to all other processors. Unlike the broadcast operation, where the same entries of data (of size count) are sent to all processes, the first count elements are send to the first process, the next count to the next and so on. MPI also provides a scatter for variable sized data (where the count size can be different for different destinations) which is called Scatterv.

The gather operation is similar to the reduce operation in that it is many-to-one with each process contributing some data to the result. The difference is that the data is not reduced but rather it is *concatenated*. If each process contributes count elements of data, then the gathering process must have space to store count multiplied by num_procs elements. There is a variable-sized data version of the gather, Gatherv in which the numbers of elements contributed per process may be different. MPI also provides Allgather and Allgatherv in which the result of the gather ends up on all the processes involved in the communication. These may be thought of as a regular Gather or Gatherv operation followed by a broadcast.

Below are the prototype signatures of the scatter and gather operations. For completeness we also give the signature of Allgatherv since we will demonstrate the use of Allgather and Allgatherv in Sect. 11.5.2.

```
void Comm::Scatter(const void* sendbuf, int sendcount,
            const MPI::Datatype& sendtype, void* recvbuf,
            int recvcount, const MPI::Datatype& recvtype,
            int root) const

void Comm::Gather(const void* sendbuf, int sendcount,
            const MPI::Datatype& sendtype, void* recvbuf,
            int recvcount, const MPI::Datatype& recvtype,
            int root) const

void Comm::Allgatherv(const void* sendbuf, int sendcount,
            const MPI::Datatype& sendtype, void* recvbuf,
            const int recvcounts[], const int displs[],
            const MPI::Datatype& recvtype) const
```

Most of the arguments in the above methods should be readily understood, since they are similar to the arguments of the previous less advanced methods. The argument root always refers the scatterer (sender) or to the gatherer (receiver). In most cases, the *types* and *counts* of the send and receive data should be identical, with the counts referring to the size of the array sent to (or received from) each process. In the variable size gather, the int array arguments recvcounts and displs are used to communicate the variable data counts and displacements for each process (so

open and close the same file; *parallel file libraries* such as MPI's MPIIO; and *separate files* where each process writes data to different places to be re-assembled later. The choice of output method is largely dependent on the data structure and size.

`recvcounts[rank]` should be equal `sendcount` for that process). The value `displs[rank]` contains the index in the gathered array `recvbuf` where the data from process `rank` should begin. There is some redundancy between the counts and displacements since one might expect the displacement of each process' data to be equal to the sum of the counts of the data from lower ranked processes. However, this redundancy allows there to be gaps in the gathered data.

11.5   Example MPI Applications

In this section, we give two examples of parallel programs written with MPI. The designs of the parallel algorithms shown here are not unique to the problems which they solve. In general, the choice of parallel algorithm depends on how it relates to an equivalent sequential algorithm (if there is one) and how the data is partitioned. One usually seeks to partition the data and computation between the processes in such a way that communication between processes is minimised and that the processes are given an equivalent amount of computational work. The task of giving the processes the same amount of work is known as *load balancing*.

However, merely giving each process a similar amount of work is no guarantee of a successful parallel algorithm if the combined computational load of parallel processes is much more than that of the sequential program, or if communication dominates the program. The measures of success in producing parallel programs are *parallel speedup* and *parallel efficiency*. The *parallel speedup* is the ratio of the time it takes to run the code sequentially on a certain problem to time it takes to run on p processes ($S_p = \frac{T_1}{T_p}$). In an ideal case, a problem can be partitioned such that it is well load balanced with minimal extra overhead, so we expect $S_p \simeq p$. Parallel efficiency scales this value by p: $E_p = \frac{T_1}{pT_p}$ so that E_p is generally in the range from 0 to 1 with 1 being the ideal value. It is uncommon, but not unusual, for a particular problem to scale in parallel such that $E_p > 1$. This fortunate situation normally arises when a given problem has memory constraints when run on a small numbers of processes and it is known as *super-linear speedup*.

11.5.1   Summation of Series

The summation of a series can be taken as an *abstraction* of a range of problems in which it is moderately easy to partition work between processes and there is minimal communication. Such problems are termed *embarrassingly parallel*. In the following example, the calculation is trivial but this case is representative of tasks which are possibly more labour intensive, such as Monte Carlo integration (see Exercise 11.4).

Consider the problem of summing a series, such as the approximation to π

$$\frac{\pi}{4} = \sum_{n=0}^{\infty} \frac{(-1)^n}{2n + 1},$$

credited to Gottfried Wilhelm Leibniz. Given that we cannot compute the sum to infinity, we approximate this summation with a finite sum from $n = 0$ to $n = max - 1$ for some value max (which may be assumed to be divisible by the number of processes, p). In dividing the max contributions between the processes evenly, we might choose to allocate this work in blocks, so that the first max/p contributions to the series go to process zero, and so on, or we might distribute in such a way as to interleave processor contributions. In the following example, the contributions are interleaved. Note that the only parallel communication needed in this code is a reduction operation, which combines the subtotals from the processes into a grand total for the entire calculation on process 0.

```cpp
#include <mpi.h>
#include <cmath>
#include <iostream>

//Program to sum Pi using Leibniz formula:
// Pi  = 4 * Sum_n ( (-1)**n/(2*n+1) )
int main(int argc, char* argv[])
{
    int max_n = 1000;
    double sum = 0;
    MPI::Init(argc, argv);

    int num_procs = MPI::COMM_WORLD.Get_size();
    int rank = MPI::COMM_WORLD.Get_rank();

    for (int n=rank; n<max_n; n+=num_procs)
    {
        double temp = 1.0/(2.0*((double)(n))+1.0);
        if (n%2 == 0) // n is even
        {
            sum += temp;
        }
        else
        {
            // n is odd
            sum -= temp;
        }
    }

    double global_sum;
    MPI::COMM_WORLD.Reduce(&sum, &global_sum, 1,
                MPI::DOUBLE, MPI::SUM, 0);
    if (rank == 0)
    {
        std::cout << "Pi is about " << 4.0*global_sum
                    << " with error " << 4.0*global_sum-M_PI
                    << "\n";
    }
    MPI::Finalize();
    return 0;
}
```

11.5.2 Parallel Linear Algebra

In this section, we give an outline of the operations required for performing parallel linear algebra operations. It is beyond the scope of this book to provide a complete parallel linear algebra library, but we outline some of the issues arising when we design such a system. A fundamental question to ask is how matrices and vectors might be partitioned across the processes. We choose to use the matrix-row partitioning (which will be described later) favoured by the PETSc library—although other parallel linear algebra systems, such as Mondriaan, use more sophisticated techniques.

We begin by discussing parallel implementation of the product between a matrix and a vector of suitable sizes. Using the matrix-row partitioning scheme, the matrix-vector product $\mathbf{v} = \mathbf{A}\mathbf{u}$ where $\mathbf{A}$ is a $N \times N$ matrix, and $\mathbf{u}$, $\mathbf{v}$ are vectors of length N, can be partitioned in such a way that the first N/p rows of matrix $\mathbf{A}$ are only known to process 0, as are the first N/p elements of the vectors $\mathbf{u}$ and $\mathbf{v}$. If we are performing a simple matrix-vector calculation using row-wise partitioning over 3 processes then it can be see from the schematic

$$
\begin{array}{c}
\text{Proc}_0 \\[6pt]
\text{Proc}_1 \\[6pt]
\text{Proc}_2
\end{array}
\left\{
\begin{pmatrix}
v_0 \\ v_1 \\ \vdots \\ \hline \vdots \\ \hline \vdots \\ v_{N-1}
\end{pmatrix}
\right.
=
\begin{pmatrix}
A_{00} & A_{01} & A_{02} & \cdots & A_{0,N-1} \\
A_{10} & A_{11} & A_{12} & \cdots & A_{1,N-1} \\
\vdots & \ddots & \ddots & \ddots & \vdots \\
\vdots & \ddots & \ddots & \ddots & \vdots \\
\vdots & \ddots & \ddots & \ddots & \vdots \\
A_{N-1,0} & A_{N-1,1} & A_{N-1,2} & \cdots & A_{N-1,N-1}
\end{pmatrix}
\begin{pmatrix}
u_0 \\ u_1 \\ \vdots \\ \vdots \\ \vdots \\ u_{N-1}
\end{pmatrix},
$$

that in order for process 0 to compute the first N/p elements of $\mathbf{v}$ it is required to know only the first N/p rows of $\mathbf{A}$ (which are held locally) and *all* the elements of $\mathbf{u}$ (most of which are not local to process 0).

More generally, in order to solve the linear system $\mathbf{A}\mathbf{x} = \mathbf{b}$ using an iterative approach (such as the conjugate gradient method described in Sect. A.2.3) there are a limited number of operations which will be needed:

- scalar-vector multiplication—an operation on locally-held data;
- vector-vector addition and subtraction—operations on locally-held data;
- a vector Euclidean norm—a sum of squares on local data, followed by a global sum of squares (a parallel reduction), followed by a square-root; and
- matrix-vector multiplication—in which, as outlined above, data from the vector must be communicated between all the processes.

We illustrate an implementation of this fashion of parallel linear algebra by giving a bare-bones working `MpiVector` class. This class contains the features listed below, which will aid building a parallel conjugate gradient solver.

- On constructing a vector of size N, the components are automatically distributed between p processes. Each process is assigned N/p elements. This division may be rounded down so that there will be a shortfall in cases where p does not divide

N. This shortfall is picked up by the top–most process. Each process holds mSize elements corresponding to indices in the range mLo $\leq i <$ mHi.

- There is an overloaded [ ] operator for accessing elements of the vector. This operator converts between a *global index* into the vector and the *local index* into the process' private data. Any out-of-range indexing trips an assertion.
- Helper methods GetHi and GetLo enable the caller to probe the range of locally held data, thus ameliorating the fact that the partitioning code is hidden from the caller which would make it easy to trip index violation assertions.
- There is a CalculateNorm method which calculates the 2-norm (see Sect. A.1.5) by calculating a local sum of squares, using reduction to sum the local sums into a global sum, and taking the square root. Note the use of Allreduce which ensures that the result of the reduction (and therefore of the norm) is available to all processes.
- There is a method UpdateGlobal for *gathering* all elements of the vector from the remote processes.

The method UpdateGlobal uses more than one gather operation as introduced in Sect. 11.4.2.4, and gathers the entire vector into private storage on every process. The first two gather operations assemble information about the number of locally held data and their displacements. These operations are here to illustrate a common use of fixed- and variable-sized gathers but they are redundant for multiple reasons: (i) the sizes and displacements are fixed in constructor and do not need to be re-calculated on every communication, (ii) the sizes and displacements are not independent— one can be calculated from the other, (iii) the algorithm for calculating sizes and displacements in the constructor is quite simple and could be repeated here.

```
1   #include <mpi.h>
2   #include <cmath>
3   #include <cassert>
4
5   class MpiVector
6   {
7   private:
8       //Store components
9       int mLo, mHi, mSize;
10      double* mData;
11      double* mGlobalData;
12  public:
13      MpiVector(int vecSize)
14      {
15          int num_procs = MPI::COMM_WORLD.Get_size();
16          int rank =  MPI::COMM_WORLD.Get_rank();
17          int ideal_local_size = vecSize/num_procs;
18
19          assert (ideal_local_size > 0);
20          mLo = ideal_local_size * rank;
21          mHi = ideal_local_size * (rank+1);
```

```
22
23        //Top processor picks up extras
24        if (rank == num_procs-1)
25        {
26            mHi = vecSize;
27        }
28        assert(mHi > mLo);
29        mData = new double[mHi - mLo];
30        mGlobalData = new double[vecSize];
31        mSize = vecSize;
32    }
33    ~MpiVector()
34    {
35        delete[] mData;
36        delete[] mGlobalData;
37    }
38
39    double& operator[](int globalIndex)
40    {
41        //Make sure that this on the local vector
42        assert(mLo<=globalIndex && globalIndex<mHi);
43        return mData[globalIndex-mLo];
44    }
45
46    int GetHi()
47    {
48        return mHi;
49    }
50
51    int GetLo()
52    {
53        return mLo;
54    }
55
56    double CalculateNorm() const
57    {
58        double local_sum = 0.0;
59        for (int i=0; i<mHi-mLo; i++)
60        {
61            local_sum += mData[i]*mData[i];
62        }
63        double global_sum;
64        MPI::COMM_WORLD.Allreduce(&local_sum, &global_sum, 1,
65                                  MPI::DOUBLE, MPI::SUM);
66        return sqrt(global_sum);
67    }
68    void UpdateGlobal()
69    {
70        int num_procs = MPI::COMM_WORLD.Get_size();
71
72        int* num_per_proc = new int[num_procs];
```

```
73     int local_size = mHi-mLo;
74     MPI::COMM_WORLD.Allgather(&local_size, 1, MPI::INT,
75                               num_per_proc, 1, MPI::INT);
76
77     int* lows_per_proc = new int[num_procs];
78     MPI::COMM_WORLD.Allgather(&mLo, 1, MPI::INT,
79                               lows_per_proc, 1, MPI::INT);
80
81     MPI::COMM_WORLD.Allgatherv(mData, local_size,
82                     MPI::DOUBLE, mGlobalData, num_per_proc,
83                     lows_per_proc, MPI::DOUBLE);
84     delete [] num_per_proc;
85     delete [] lows_per_proc;
86   }
87 };
```

```
1  #include <iostream>
2  #include <mpi.h>
3  #include "MpiVector.hpp"
4
5  int main(int argc, char* argv[])
6  {
7     MPI::Init(argc, argv);
8     MpiVector all_ones(9);
9     std::cout << "Local has [" << all_ones.GetLo() <<
10                  ", " << all_ones.GetHi() << ")\n";
11    for (int i=all_ones.GetLo(); i<all_ones.GetHi(); i++)
12    {
13       all_ones[i] = 1.0;
14    }
15    assert( fabs(all_ones.CalculateNorm()-3.0) < 1.0e-6 );
16
17    all_ones.UpdateGlobal();
18    MPI::Finalize();
19    return 0;
20 }
```

11.6   Tips: Debugging a Parallel Program

We have discussed debugging sequential code in Sects. 1.7 and 7.7. Message passing clearly introduces the potential for different errors to be inserted into your code. We discuss some methods for debugging parallel programs below.

11.6.1 Tip 1: Make an Abstract Program

As with sequential programming, it is very rare for a programmer to begin building a parallel program from scratch. In many cases, you may be given a sequential program which has been written by someone else, or you may be starting from your own program. At such times, it is hard to see the communication patterns underlying the parallel code—they can easily get lost in the details of the calculations.

Our advice is to first take the time to design a rough idea of the communication patterns needed in your new parallel program, and then start afresh. Write a simplified *abstract program* which concentrates on the communication, but neglects the main calculation. This will give you the opportunity to ensure the safe working of the parallel communication in the absence of details of the particulars. Once the message passing is working correctly, it can easily be integrated into the main code.

11.6.2 Tip 2: Datatype Mismatch

In the following code, copied incorrectly from Listing 11.2, the process 0 block has been amended so that the type of the data is now int and the message is sent as MPI::INT. However, this change has not been reflected in the code for the receiving process where the message is received as MPI::DOUBLE.

```
1   int tag = 30;
2   if (MPI::COMM_WORLD.Get_rank() == 0)
3   {
4       //Specific send code for process 0
5       int send_buffer[2] = {100, 200};
6       MPI::COMM_WORLD.Send(send_buffer, 2,
7                            MPI::INT, 1, tag);
8   }
9   if (MPI::COMM_WORLD.Get_rank() == 1)
10  {
11      //Specific receive code for process 1
12      double recv_buffer[2] = {0.0, 0.0};
13      MPI::COMM_WORLD.Recv(recv_buffer, 2, MPI::DOUBLE,
14                           MPI::ANY_SOURCE, MPI::ANY_TAG);
15      std::cout << recv_buffer[0] << "\n";
16      std::cout << recv_buffer[1] << "\n";
17  }
```

The message passing in this program may work correctly—in terms of the communication pattern—but the data received on process 1 will probably be incorrect. This may be because of mismatches in the *size* of the data (on most architectures int uses 32 bits whereas double uses 64 bits) or it may be due to errors in the *conversion* of the data.

Problems where message data types (or sizes) do not match can be hard to see especially when the send and receive components are in separate methods or in separate files.

11.6.3   Tip 3: Intermittent Deadlock

Deadlock is the technical term for the situation in which all processes are waiting for some event to happen before proceeding but no process can supply that event because they are waiting for another process. This situation is illustrated simply by four cars arriving simultaneously at a junction where the traffic signals have failed: with nothing to tell them how to proceed all four drivers play safe and wait for someone else to make the first move. In most cases, it is possible to find code which causes deadlock by heavily instrumenting the program, that is, by printing out lots of information and flushing the output. We will deliberately induce deadlock in Exercise 11.2 by never receiving any sent messages so that eventually the sender is not able to proceed because it is unable to send any more messages.

Problems involving *intermittent deadlock* are harder to diagnose. These are situations where the program deadlocks on some runs of the code but runs normally on others. Perhaps the program runs without encountering problems with some trivial example test input, but when it is fed with the real-life input it then deadlocks. When this happens, it is an indication that the problem is to do with the size or timing of messages. In Exercise 11.2 we demonstrate that small amounts of data can be buffered—which hides the fact that a non-buffered blocking send would produce deadlock—but large amounts of data cannot be buffered. In other words, for a given program there may be sizes and timings of messages where deadlock happens, and some where it does not happen.

In such situations, a good strategy is to concentrate on those situations most likely to deadlock. We make our program less efficient and more likely to deadlock by removing buffering and asynchronous messages: replacing all instances of Send with Ssend. Once all message passing is synchronous it is likely that the intermittent deadlock has become predictable deadlock, allowing us to identify the problem and debug the code. A program can also be made "more synchronous" by splitting calculation steps up with barriers. The program can later be made more efficient as necessary.

11.6.4   Tip 4: Almost Collective Communication

It is common to treat process zero as a "master process", orchestrating the tasks of the other processes, reducing data for output to the screen, and gathering information from all processes for output to a single file. In these circumstances, it is usual to have some blocks of code or some methods which are only executed by the master process and some which are only executed by the "slave processes".

In Sect. 11.4.2.4, we gave the example of an output pattern in which all data was *concentrated* onto a single process before being written to disk. In this case, process zero may execute a block of code consisting of receives and writes to disk via an `ofstream`, whereas the other processes execute a block consisting of the matching send commands. When debugging parallel code, it is usually a good idea to add barriers in order to break the program into manageable sections. However, if we were to add barriers into the slave processes' block of sending code, this would be a recipe for instant deadlock. Since all processes *except one* are executing this code, then any collective communication on `MPI::COMM_WORLD` cannot complete. If collective communication is necessary in this code, then a new communication group (including all processes in `MPI::COMM_WORLD` except rank zero) must be created. New communication groups can be created using relevant MPI functions such as `MPI_Comm_split` (see MPI documentation for more details).

11.7 Exercises

11.1 Amend the `MpiHelloWorld` program in Listing 11.1 so that the processes print in reverse rank order. You can do this with a down-loop over processes and a barrier. Beware that if your implementation of MPI buffers output then you might not be able to verify that your process is working correctly!

Assuming that your loop for output is correct, now modify it to do *round robin* file output. Instead of writing process ranks to `std::cout` each process in turn should: open a named file, write the rank information to it and close the file. The second process to write (and those subsequent) should not open the file until the previous process has closed it and should open the file in append mode (see Sect. 3.2).

Investigate the `MPI::Wtime` method (which returns a high-precision time, with units of seconds, since some fixed point of time in the past) and use it to time the program on each process. Use `Reduce` to compute the average duration of the program over all processes.

11.2 The MPI standard allows the `Send` library call to behave either like a buffered send or like a blocking send. In practice, all implementations of the MPI standard treat `Send` the same way. If the message is small enough (and there is space), then it is copied into a private buffer, and the MPI library is delegated to ensure that the message is delivered and the program flow continues—similar to `Bsend`. If the message is large (or if that private buffer is full), then delivery of the message must wait until the recipient is ready for it, so the program flow waits—similar to `Ssend`.

Write an MPI program where the master process has one loop which attempts to send larger messages each time, and then prints how big the message was. We suggest that you double the size of the message on each iteration. All other processes should do nothing. We suggest that you have an array of length *at least* a million items, to make sure that there is always something to be sent. Eventually you should observe deadlock.

11.3 Write an MPI code following the instructions below. This code is to be executed with only two processes, and tests the use of MPI for transferring vectors of data between processes.

- Define an array V[10][10] to store the entries of a 10 × 10 matrix. The process with rank 0 initialises its copy of the array to

$$V[row][col] = 10*row+col,$$

  while the process with rank 1 initialises its copy of the array to

$$V[row][col] = 100+10*row+col.$$

  This choice provides a convenient way of identifying, from the value of the entry of V, where it has come from in the original arrays, and from which process: the three-digit value xyz will be row y, column z from process x.
- Transfer the data stored in the first row of the matrix stored by process 0 into the corresponding positions in the matrix stored by process 1. This involves process 0 sending the data using Send, and process 1 receiving the data using Recv. One way of doing this on the sending side is to first copy the data into a buffer vector of suitable length and then send this vector. Similarly, on the receiving side receive it into a buffer vector of suitable length and then copy it into the appropriate part of V.
- Print out the contents of the array V stored by process 1 to check that you have correctly sent the data.
- Repeat the transfer of the first row of data without copying into a buffer on the sender or copying from a buffer on the receiver.
- Repeat the transfer of data sending both the row with index 5 and the row with index 8 between the processes.
- Transfer the first *column* of data between the processes.

11.4 The aim of this exercise is to get you started on writing algorithms with collective communications. The exercise asks you to develop a parallel algorithm for calculating an approximation to π using Monte Carlo integration.

Suppose we want to approximate the integral

$$\int_a^b f(x)\,dx,$$

where $f(x)$ is a continuous function defined at all points in the closed interval $a \le x \le b$. If X_i, $i = 0, 1, 2, \ldots, N-1$ are independent random variables uniformly distributed on the interval $a \le x \le b$, where N is sufficiently large, then Monte Carlo integration allows us to approximate the integral by

$$\int_a^b f(x)\,dx \approx \frac{b-a}{N}\sum_{i=0}^{N-1} f(X_i).$$

Noting that

$$\pi = 4 \int_0^1 \frac{1}{1+x^2} dx,$$

we will use Monte Carlo integration to estimate π through approximating the integral on the right-hand side of this equation. Sequential code for this is given below.

The random numbers are generated through the random number generator `rand` (line 18), and seeded through `srand` (line 11). The random number generator requires the `cstdlib` header to be included. The random number generator is seeded differently on every run: in this exercise you will develop this code to run on a distributed memory machine through use of MPI statements, and you don't want a set of parallel computers to all work on the same set of "random" numbers.

```cpp
1    // Compute pi using Monte Carlo integration
2    // of 1/(1+x*x) on the interval 0<=x<=1
3    #include <cmath>
4    #include <cstdlib>
5    #include <iostream>
6    #include <unistd.h> //For getpid()
7
8    int main(int argc, char* argv[])
9    {
10       // seed random number generator
11       srand(getpid());
12       int n_points = 1000000;
13
14       double sum = 0;
15       for (int i=0; i<n_points; i++)
16       {
17          // generate a random number on the interval 0<=x<=1
18          double x = rand()/((double)(RAND_MAX));
19          double f = 1.0/(1.0+x*x);
20          sum += f;
21       }
22       double pi = 4.0*(sum/((double)(n_points)));
23       std::cout << "Pi is approximately " << pi
24                 << " with error " << pi-M_PI << "\n";
25
26       return 0;
27    }
```

Compile the program and it should print out an answer similar to

```
Pi is approximately 3.141782 with error 0.000355562
```

In the exercises below, we will add MPI function calls to enable this code to be run in parallel.

1. Add MPI function calls so that n_points function evaluations are performed on each of the MPI processes.
2. Estimate π through reducing the result of function evaluations (sum) from each processor to a global sum on process 0 and scaling appropriately. This is similar to the summation of a series in Sect. 11.5.1.
3. Amend the code so that process 0 selects a value of n_points for each of the processes at the beginning program. Pass these values out in a scatter operation.

11.5 Write classes to enable parallel linear algebra based on the row-wise matrix partitioning—and the MpiVector class—given Sect. 11.5.2. Your goal for this exercise should be to perform a matrix-vector multiplication in parallel.

1. Add as much functionality and overloaded operators from the Vector class given in Sect. 10.1 to MpiVector as you wish. Include any improvements which you may have made to Vector as part of Exercise 10.2.
2. The MpiVector constructor contains an assertion that the ideal local size (number of local vector elements) should be nonzero. This guards against the case when the number of processes is larger than vecSize, in which case the current code in the constructor would assign the entire vector to the top-most process. Fix this situation so that when there are fewer vector elements than processes every process is assigned either one or zero elements.
3. Make it possible to set elements on remote processes. A suitable scheme would be to construct the vector in "set up" mode, during which requests to add values to remote elements are stored for later. A user is able to call a method FinishSetUp which communicates the stored data between processes, puts the vector in a "usable" mode and bars future attempts to set remote data.
4. Remove some of the redundant calculations performed by UpdateGlobal mentioned in Sect. 10.1.
5. Write an output method which uses UpdateGlobal such that one process is able to print the entire vector to screen or to file.
6. The UpdateGlobal method relies on memory for the private data member mGlobalData being allocated in the constructor. Since the mGlobalData is only required for output or for a matrix-vector product, the memory for mGlobalData ought to be allocated on demand. Make sure that there is also a method for de-allocating this memory when it is no longer needed.
7. Write an MpiMatrix class using the scheme outlined in Sect. 10.1. It is important that you treat the partition on the number of matrix rows in exactly the same way as the vector partition, so that local sizes are always compatible. Perform a matrix-vector multiplication in parallel and output the solution.

Designing Object-Oriented Numerical Libraries

Having developed classes that underpin linear algebra operations in Chap. 10 we now demonstrate how to construct object-oriented libraries for scientific computing applications that utilise the functionality of these classes. We use the specific example of developing a library that uses the finite difference method to solve boundary value, second order differential equations.

We begin by developing a library for problems in one spatial dimension that are linear, constant coefficient, second order, boundary value ordinary differential equations. That is, equations of the form

$$A\frac{d^2u}{dx^2} + B\frac{du}{dx} + Cu = f(x), \qquad X_0 < x < X_1, \tag{12.1}$$

where A ($\neq 0$), B, C, X_0, X_1 (with $X_0 < X_1$) are given constants, $f(x)$ is a given function, and suitable boundary conditions are given at $x = X_0$ and $x = X_1$. We choose to use the finite difference method to underpin the library as this method for calculating the numerical solution of differential equations is the simplest to explain, and a method that many readers will be familiar with. This allows us to focus on the *implementation* of this method, without a need to explain more technical aspects of the method from a mathematical viewpoint as would be the case with more sophisticated techniques such as the finite element method. Having discussed how to develop a library for this class of equations we conclude this chapter by briefly touching upon how a library for computing the numerical solution of Poisson's equation may be constructed. For ease of explanation, we limit ourselves to a two-dimensional rectangular domain, and apply only Dirichlet boundary conditions, that is, the following partial differential equation:

$$\frac{\partial^2 u}{\partial x^2} + \frac{\partial^2 u}{\partial y^2} = f(x, y), \qquad X_0 < x < X_1, Y_0 < y < Y_1,$$

© Springer International Publishing AG, part of Springer Nature 2017
J. Pitt-Francis and J. Whiteley, *Guide to Scientific Computing in C++*, Undergraduate Topics in Computer Science,
https://doi.org/10.1007/978-3-319-73132-2_12

where X_0, X_1, Y_0, Y_1 are specified constants with $X_0 < X_1, Y_0 < Y_1, f(x, y)$ is a specified function, and u is specified at each point on the boundary. As partial differential equations may be beyond the mathematical scope of some readers, this section is entirely self-contained: the remainder of this chapter may be read independently of the material in Sect. 12.3.

The emphasis of this chapter is to explain the object-oriented structure that may be used when developing a library for solving differential equations. We describe the functionality required from the classes that we use, but give very little detail on the implementation of these classes: implementation of the ideas presented uses C++ techniques introduced in earlier chapters, and is the focus of the exercises at the end of the chapter. The mathematical theory of the finite difference method is not discussed in much detail. Readers unfamiliar with this technique should consult a suitable text such as Iserles [1], Kreyszig [2], or Süli and Mayers [3].

12.1 Developing the Library for Ordinary Differential Equations

When developing software, it is useful to know precisely what type of problems are to be solved using this software. We therefore begin by defining two exemplar model problems that contain all features commonly seen in linear, constant coefficient, boundary value ordinary differential equations. We then explain the mathematical theory behind the finite difference method for these boundary value problems, before concluding this section by explaining how to utilise the theory when developing the library.

12.1.1 Model Problems

We use two example model problems to motivate the development of the library. These model problems have a known solution and can therefore be used to give some verification of the correctness of the output of the library. The first model problem is very simple, whilst the second model problem is more complicated and uses all the features that we will include in our library for ordinary differential equations.

Model Problem 1. The first model problem is the following boundary value problem:

$$\frac{d^2 u}{dx^2} = -1, \quad 0 < x < 1,$$

$$u = 0, \quad \text{at } x = 0,$$

$$u = 0, \quad \text{at } x = 1.$$

This problem has solution

$$u(x) = \frac{1}{2}x(1-x).$$

This is a very simple problem—we have the minimal number of terms in the differential equation, and only very simple Dirichlet (i.e., non-derivative) boundary conditions.

Model Problem 2.   The second model problem is a more complicated differential equation, with one Dirichlet boundary condition, and one Neumann (derivative) boundary condition. This model problem satisfies the following equation and boundary conditions:

$$\frac{d^2u}{dx^2} + 3\frac{du}{dx} - 4u = 34\sin x, \qquad 0 < x < \pi,$$

$$\frac{du}{dx} = -5, \qquad \text{at } x = 0,$$

$$u = 4, \qquad \text{at } x = \pi.$$

This differential equation has solution

$$u = \frac{4e^x + e^{-4x}}{4e^\pi + e^{-4\pi}} - 5\sin x - 3\cos x.$$

12.1.2 Finite Difference Approximation to Derivatives

We now define the notation used for the finite difference approximations to the first and second derivative of a function of one variable. Where we define a derivative at N distinct points, we will denote these points using subscripts starting at 1 and ending at N for consistency with the overloaded parenthesis operators used when writing the classes of vectors and matrices developed in Chap. 10.

Let us suppose that a function u is defined on the interval $X_0 \le x \le X_1$. Suppose further that there is a collection of points x_i, $i = 1, 2, \ldots, N$, that satisfy

$$x_1 = X_0,$$
$$x_1 < x_2 < x_3 < \cdots < x_N,$$
$$x_N = X_1.$$

We will refer to the points $x_1, x_2, \ldots, x_N$ as the *finite difference grid*, and the individual points as *nodes*. The nodes x_1 and x_N are referred to as the *boundary nodes* of

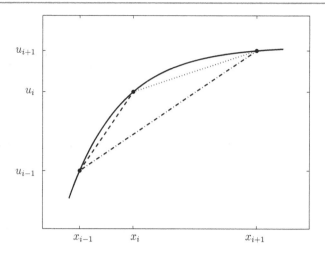

Fig. 12.1 Backward finite difference (*broken line*), forward finite difference (*dotted line*), and central finite difference (*dot-dashed line*) approximations to the first derivative of the function represented by *the solid line* at the point $x = x_i$

Table 12.1 Numerical finite difference approximations to the first derivative at $x = x_i$

Type	Formula	Range
Backward	$(u_i - u_{i-1}) / (x_i - x_{i-1})$	$i = 2, 3, \ldots, N$
Forward	$(u_{i+1} - u_i) / (x_{i+1} - x_i)$	$i = 1, 2, \ldots, N - 1$
Central	$(u_{i+1} - u_{i-1}) / (x_{i+1} - x_{i-1})$	$i = 2, 3, \ldots, N - 1$

the finite difference grid, whilst all other points are referred to as *interior nodes*. We may evaluate the function u at each node x_i, $i = 1, 2, \ldots, N$, which we denote by u_i:

$$u_i = u(x_i).$$

The first derivative of a function at a given node may be thought of as being the "slope" of the function at that point: i.e. the ratio of the change in u to the change in x. In Fig. 12.1 we motivate three different approximations to the first derivative at $x = x_i$ which are defined in Table 12.1. Note that not all of these approximations are defined at the boundary nodes of the finite difference grid, that is, at $x = x_1$ and $x = x_N$.

A numerical approximation to the second derivative, not defined at the boundary nodes of the finite difference grid, $x = x_1$ and $x = x_N$, is

$$\frac{2}{x_{i+1} - x_{i-1}} \left(\frac{u_{i+1} - u_i}{x_{i+1} - x_i} - \frac{u_i - u_{i-1}}{x_i - x_{i-1}} \right), \qquad i = 2, 3, \ldots, N - 1,$$

which may be written

$$\alpha_i u_{i-1} + \beta_i u_i + \gamma_i u_{i+1}, \qquad i = 2, 3, \ldots, N - 1, \tag{12.2}$$

where

$$\alpha_i = \frac{2}{(x_{i+1} - x_{i-1})(x_i - x_{i-1})}, \tag{12.3}$$

$$\beta_i = -\frac{2}{(x_{i+1} - x_i)(x_i - x_{i-1})}, \tag{12.4}$$

$$\gamma_i = \frac{2}{(x_{i+1} - x_{i-1})(x_{i+1} - x_i)}. \tag{12.5}$$

This approximation to the second derivative follows from Taylor series expansions: see, for example, Kreyszig [2]. We note that when there is a uniform spacing between the nodes, that is, $x_{i+1} - x_i = h, i = 1, 2, 3, \ldots, N - 1$, for some constant h, then the approximation to the second derivative given in Eq. (12.2) may be simplified to the more familiar formula

$$\frac{u_{i+1} - 2u_i + u_{i-1}}{h^2}.$$

When developing our classes we will use the approximation given in Eq. (12.2) as it allows more generality.

12.1.3 Application of Finite Difference Methods to Boundary Value Problems

We now explain how the finite difference approximations given in Sect. 12.1.2 may be used to calculate a numerical solution of the model problems given in Sect. 12.1.1. For both problems we use the finite difference grid with N nodes described in Sect. 12.1.2. There are therefore N unknown values of u_i to determine. We will demonstrate how to set up a linear system of size N that allows us to calculate these values.

12.1.3.1 Model Problem 1

Substituting the approximation to second derivative given by Eq. (12.2) into the differential equation at the interior nodes of the finite difference grid yields

$$\alpha_i u_{i-1} + \beta_i u_i + \gamma_i u_{i+1} = -1, \qquad i = 2, 3, \ldots, N - 1. \tag{12.6}$$

The boundary conditions imply that

$$u_1 = u_N = 0. \tag{12.7}$$

Equations (12.6) and (12.7) may be combined and written as the linear system $\mathbf{Au} = \mathbf{b}$, where $\mathbf{A}$ is a $N \times N$ matrix, and $\mathbf{u}$ and $\mathbf{b}$ are vectors of length N. The entries of $\mathbf{A}$, $\mathbf{u}$ and $\mathbf{b}$ are then given by

$$
\mathbf{A} = \begin{pmatrix}
1 & 0 & 0 & \cdots & 0 & 0 & 0 \\
\alpha_2 & \beta_2 & \gamma_2 & \cdots & 0 & 0 & 0 \\
0 & \alpha_3 & \beta_3 & \cdots & 0 & 0 & 0 \\
0 & 0 & \alpha_4 & \cdots & 0 & 0 & 0 \\
\vdots & \vdots & \vdots & \ddots & \vdots & \vdots & \vdots \\
0 & 0 & 0 & \cdots & \alpha_{N-1} & \beta_{N-1} & \gamma_{N-1} \\
0 & 0 & 0 & \cdots & 0 & 0 & 1
\end{pmatrix},
$$

$$
\mathbf{u} = \begin{pmatrix}
u_1 \\ u_2 \\ u_3 \\ u_4 \\ \vdots \\ u_{N-2} \\ u_{N-1} \\ u_N
\end{pmatrix}, \quad
\mathbf{b} = \begin{pmatrix}
0 \\ -1 \\ -1 \\ -1 \\ \vdots \\ -1 \\ -1 \\ 0
\end{pmatrix}.
$$

Now that we have written the model problem as a linear system, we may use the methods associated with the vector, matrix and linear system classes to solve this system and calculate the values of u_i, $i = 1, 2, \ldots, N$.

12.1.3.2 Model Problem 2

We now write model problem 2 in matrix form. At the interior nodes of the finite difference grid, we use a central approximation to the first derivative, as defined in Table 12.1, and the approximation to the second derivative given by Eq. (12.2). The differential equation may then be approximated by, for $i = 2, 3, \ldots, N-1$,

$$
\left(\alpha_i - \frac{3}{x_{i+1} - x_{i-1}} \right) u_{i-1} + (\beta_i - 4) u_i + \left(\gamma_i + \frac{3}{x_{i+1} - x_{i-1}} \right) u_{i+1} = 34 \sin x_i.
$$

$$(12.8)$$

The boundary condition at $x = \pi$ may be implemented in the same way as the Dirichlet boundary conditions in model problem 1, that is, we write

$$
u_N = 4. \tag{12.9}
$$

The Neumann (derivative) boundary condition at $x = 0$ requires a bit more thought. We see from Table 12.1 that the only one of these approximations to the first derivative that is defined at the node x_1 is the forward approximation. We therefore use this approximation and implement this boundary condition by setting

$$-\frac{1}{x_2 - x_1}u_1 + \frac{1}{x_2 - x_1}u_2 = -5. \tag{12.10}$$

Defining, for $i = 2, 3, \ldots, N - 1$, the quantities $\hat{\alpha}_i, \hat{\beta}_i, \hat{\gamma}_i$:

$$\hat{\alpha}_i = \alpha_i - \frac{3}{x_{i+1} - x_{i-1}},$$

$$\hat{\beta}_i = \beta_i - 4,$$

$$\hat{\gamma}_i = \gamma_i + \frac{3}{x_{i+1} - x_{i-1}},$$

we may write Eqs. (12.8)–(12.10) as the linear system $\mathbf{Au} = \mathbf{b}$, where the entries of $\mathbf{A}$ and $\mathbf{b}$ are given by

$$\mathbf{A} = \begin{pmatrix} -1/(x_2 - x_1) & 1/(x_2 - x_1) & 0 & \cdots & 0 & 0 & 0 \\ \hat{\alpha}_2 & \hat{\beta}_2 & \hat{\gamma}_2 & \cdots & 0 & 0 & 0 \\ 0 & \hat{\alpha}_3 & \hat{\beta}_3 & \cdots & 0 & 0 & 0 \\ 0 & 0 & \hat{\alpha}_4 & \cdots & 0 & 0 & 0 \\ \vdots & \vdots & \vdots & \ddots & \vdots & \vdots & \vdots \\ 0 & 0 & 0 & \cdots & \hat{\alpha}_{N-1} & \hat{\beta}_{N-1} & \hat{\gamma}_{N-1} \\ 0 & 0 & 0 & \cdots & 0 & 0 & 1 \end{pmatrix},$$

$$\mathbf{u} = \begin{pmatrix} u_1 \\ u_2 \\ u_3 \\ u_4 \\ \vdots \\ u_{N-2} \\ u_{N-1} \\ u_N \end{pmatrix}, \quad \mathbf{b} = \begin{pmatrix} -5 \\ 34\sin(x_2) \\ 34\sin(x_3) \\ 34\sin(x_4) \\ \vdots \\ 34\sin(x_{N-2}) \\ 34\sin(x_{N-1}) \\ 4 \end{pmatrix}.$$

As with model problem 1 we may now use the linear system class already written to solve this linear system.

12.1.4 Concluding Remarks on Boundary Value Problems in One Dimension

We have now explained how to write the finite difference approximation to a linear, constant coefficient, second order boundary value problem in matrix notation, thus allowing the classes of vectors, matrices and linear systems developed in Chap. 10 to be used to calculate the finite difference approximation. In the next section, we wil describe an object-oriented structure that allows a very general library for solving such problems to be developed. We should, however, discuss the limitations of this library.

Suppose we want to solve the following equation:

$$\frac{d^2u}{dx^2} + u = 0, \quad 0 < x < 2\pi,$$
$$u = 0, \quad \text{at } x = 0,$$
$$u = 0, \quad \text{at } x = 2\pi.$$

This has solution $u = \sin x$, and it may be thought that the library we are writing may be used to solve this problem. However $u = A \sin x$, where A is any constant value satisfies the differential equation and both boundary conditions: that is, the solution is not unique.

The equation above has a non-unique solution. It is also possible that an equation of the form Eq. (12.1) has no solution. For example, consider the equation

$$\frac{d^2u}{dx^2} + u = 0, \quad 0 < x < 2\pi,$$
$$u = 1, \quad \text{at } x = 0,$$
$$u = 4, \quad \text{at } x = 2\pi.$$

It can be shown that this equation, together with these boundary conditions, has no solution.

Proof of existence and uniqueness of solutions to boundary value differential equations is beyond the scope of this book. Nevertheless, the reader should be aware when using this library that some equations have solutions that are not unique, and solutions do not exist for other equations.

12.2 Designing a Library for Solving Boundary Value Problems

To calculate a numerical solution of the boundary value ordinary differential equations discussed above, we may specify the problem by specifying individually: (i) the ordinary differential equation and the interval on which the solution is valid; (ii) the boundary conditions; and (iii) the finite difference grid. Classes will be written for these three entities, called `SecondOrderOde`, `BoundaryConditions` and `FiniteDifferenceGrid`. These will then all be members of a class `BvpOde` that encapsulates a boundary value ordinary differential equation, and contains all the functionality required for the numerical solution of the differential equation. We now discuss the individual classes.

12.2.1 The Class `SecondOrderOde`

To specify the ordinary differential equation, we need to specify the coefficients on the left-hand side of Eq. (12.1), the function on the right-hand side of this equation, and the interval on which the equation is valid. These will all be made members of the class `SecondOrderOde`. To ensure that all of these are specified, we will only allow a user to use a constructor that specifies all of these members. In the exercises at the end of this chapter, we will discuss developing other constructors. A header file for this class is given below.

Listing 12.1 SecondOrderOde.hpp

```cpp
#ifndef SECONDORDERODEHEADERDEF
#define SECONDORDERODEHEADERDEF

class SecondOrderOde
{
   // The boundary value class is able to
   // access the coefficients etc. of this equation
   friend class BvpOde;
private:
   // Coefficients on LHS of ODE
   double mCoeffOfUxx;
   double mCoeffOfUx;
   double mCoeffOfU;
   // Function on RHS of ODE
   double (*mpRhsFunc)(double x);

   // Interval for domain
   double mXmin;
   double mXmax;
public:
   SecondOrderOde(double coeffUxx, double coeffUx,
                  double coeffU,
                  double (*righthandSide)(double),
                  double xMinimum, double xMaximum)
   {
      mCoeffOfUxx = coeffUxx;
      mCoeffOfUx = coeffUx;
      mCoeffOfU = coeffU;
      mpRhsFunc = righthandSide;
      mXmin = xMinimum;
      mXmax = xMaximum;
   }
};

#endif
```

12.2.2 The Class `BoundaryConditions`

On the left boundary, we may specify either the value of the function u (a left Dirichle
boundary condition), or the derivative du/dx (a left Neumann boundary condition)
It is important to note that there must be *either* a left Dirichlet boundary conditior
or a left Neumann boundary condition: we must have one of these boundary conditions but we cannot have both. Similarly, on the right boundary we must have
either a right Dirichlet boundary condition *or* a right Neumann boundary condition
In the class `BoundaryConditions`, we will declare class members `mLhsBc-`
`IsDirichlet, mRhsBcIsDirichlet, mLhsBcIsNeumann, mRhsBcIs-`
`Neumann` that are Boolean variables, thus allowing us to check that we have precisely one boundary condition on the left–hand boundary, and precisely one boundary condition on the right boundary. The default constructor should be overridden
to set these variables to the value "`false`" in the absence of any other instruction
Whatever type of boundary conditions are set, values for these are needed at either
end of the interval. These class members are called `mLhsBcValue` and `mRhs-`
`BcValue`. Finally, we require methods to set these values, and set the appropriate
Boolean variable to the value "`true`". The method `SetLhsDirichletBc` takes a
double precision floating point variable as input. It sets the member variable `mLhs-`
`BcValue` to this input, and sets the Boolean variable `mLhsBcIsDirichlet` tc
the value `true`. The methods `SetRhsDirichletBc, SetLhsNeumannBc` and
`SetRhsNeumannBc` perform similar tasks.

The header file `BoundaryConditions.hpp` is shown below.

Listing 12.2 BoundaryConditions.hpp

```
1   #ifndef BOUNDARYCONDITIONSHEADERDEF
2   #define BOUNDARYCONDITIONSHEADERDEF
3
4   class BoundaryConditions
5   {
6   public:
7      // The boundary value class is able to
8      // access the coefficients etc. of this equation
9      friend class BvpOde;
10  private:
11     bool mLhsBcIsDirichlet;
12     bool mRhsBcIsDirichlet;
13     bool mLhsBcIsNeumann;
14     bool mRhsBcIsNeumann;
15     double mLhsBcValue;
16     double mRhsBcValue;
17  public:
18     BoundaryConditions();
19     void SetLhsDirichletBc(double lhsValue);
20     void SetRhsDirichletBc(double rhsValue);
21     void SetLhsNeumannBc(double lhsDerivValue);
22     void SetRhsNeumannBc(double rhsDerivValue);
23  };
24
25  #endif
```

12.2.3 The Class `FiniteDifferenceGrid`

The finite difference grid requires access to the interval on which the equation is valid, given in the class `SecondOrderOde`. To create a uniform grid, we also need specification of the number of nodes. To ensure that the number of nodes is specified, we only allow use of a constructor that sets this through a constructor argument. A vector of uniformly spaced nodes can then be generated. We create a class `Node` that stores the coordinate of each node. Header files for the classes `FiniteDifferenceGrid` and `Node` are given below.

Listing 12.3 `Node.hpp`

```
1  #ifndef NODEHEADERDEF
2  #define NODEHEADERDEF
3
4  class Node
5  {
6  public:
7      double coordinate;
8  };
9
10 #endif
```

Listing 12.4 `FiniteDifferenceGrid.hpp`

```
1  #ifndef FINITEDIFFERENCEGRIDHEADERDEF
2  #define FINITEDIFFERENCEGRIDHEADERDEF
3  #include <vector>
4  #include "Node.hpp"
5
6  class FiniteDifferenceGrid
7  {
8  public:
9      // The boundary value class is able to
10     // access the nodes
11     friend class BvpOde;
12 private:
13     std::vector<Node> mNodes;
14 public:
15     FiniteDifferenceGrid(int numNodes, double xMin,
16                          double xMax);
17 };
18
19 #endif
```

12.2.4 The Class BvpOde

Now we have described the classes SecondOrderOde, BoundaryConditions and FiniteDifferenceGrid we may develop the class BvpOde. We only allow this class to be instantiated through a constructor that specifies: (i) an instance of the class SecondOrderOde; (ii) an instance of the class BoundaryConditions and (iii) the number of nodes to be used in the finite difference grid. Once these entities have been specified we then create an instance of the class FiniteDifference-Grid, a vector that will contain the solution, a vector that will be on the right–hand side of a linear system, and a matrix associated with the linear system. Methods will then be written to populate both the matrix and the vector associated with the linear system, and to apply the boundary conditions, as discussed in Sect. 12.1.3. Finally methods will be written to solve the linear system, and to write the solution to file. A header file BvpOde.hpp is given below.

Listing 12.5 BvpOde.hpp

```
 1  #ifndef BVPODEHEADERDEF
 2  #define BVPODEHEADERDEF
 3
 4  #include <string>
 5  #include "Matrix.hpp"
 6  #include "Vector.hpp"
 7  #include "LinearSystem.hpp"
 8  #include "FiniteDifferenceGrid.hpp"
 9  #include "SecondOrderOde.hpp"
10  #include "BoundaryConditions.hpp"
11
12  class BvpOde
13  {
14  private:
15     // Only allow instance to be created from a PDE, boundary
16     // conditions, and number of nodes in the mesh (the
17     // copy constructor is private)
18     BvpOde(const BvpOde& otherBvpOde){}
19
20     // Number of nodes in the grid, and a pointer to a grid
21     int mNumNodes;
22     FiniteDifferenceGrid* mpGrid;
23
24     // Pointer to instance of an ODE
25     SecondOrderOde* mpOde;
26
27     // Pointer to an instance of boundary conditions
28     BoundaryConditions* mpBconds;
29
30     // Vector for solution to unknowns
31     Vector* mpSolVec;
32
33     // Right-hand side vector
34     Vector* mpRhsVec;
```

```
35
36     // Matrix for linear system
37     Matrix* mpLhsMat;
38
39     // Linear system that arises
40     LinearSystem* mpLinearSystem;
41
42     // Allow user to specify the output file or
43     // use a default name
44     std::string mFilename;
45
46
47     // Methods for setting up linear system and solving it
48     void PopulateMatrix();
49     void PopulateVector();
50     void ApplyBoundaryConditions();
51
52 public:
53     // Sole constructor
54     BvpOde(SecondOrderOde* pOde, BoundaryConditions* pBcs,
55                                    int numNodes);
56
57     // As memory is dynamically allocated the destructor
58     // is overridden
59     ~BvpOde();
60
61     void SetFilename(const std::string& name)
62     {
63         mFilename = name;
64     }
65     void Solve();
66     void WriteSolutionFile();
67 };
68
69 #endif
```

12.2.5 Using the Class BvpOde

When using the classes introduced above, we would like to write code such as that in Listing 12.6 to calculate a numerical solution of the model problems given in Sect. 12.1.1. This will form the basis for the exercises at the end of this chapter.

Listing 12.6 `Driver.cpp` for testing the code in Sect. 12.2 on the model problems discussed in Sect. 12.1.1

```cpp
#include <cmath>
#include <string>
#include "BvpOde.hpp"

double model_prob_1_rhs(double x){return 1.0;}
double model_prob_2_rhs(double x){return 34.0*sin(x);}

int main(int argc, char* argv[])
{
    SecondOrderOde ode_mp1(-1.0, 0.0, 0.0,
                           model_prob_1_rhs,
                           0.0, 1.0);
    BoundaryConditions bc_mp1;
    bc_mp1.SetLhsDirichletBc(0.0);
    bc_mp1.SetRhsDirichletBc(0.0);

    BvpOde bvpode_mp1(&ode_mp1, &bc_mp1, 101);
    bvpode_mp1.SetFilename("model_problem_results1.dat");
    bvpode_mp1.Solve();

    SecondOrderOde ode_mp2(1.0, 3.0, -4.0,
                           model_prob_2_rhs,
                           0.0, M_PI);
    BoundaryConditions bc_mp2;
    bc_mp2.SetLhsNeumannBc(-5.0);
    bc_mp2.SetRhsDirichletBc(4.0);

    BvpOde bvpode_mp2(&ode_mp2, &bc_mp2, 1001);
    bvpode_mp2.SetFilename("model_problem_results2.dat");
    bvpode_mp2.Solve();

    return 0;
}
```

12.3   Extending the Library to Two Dimensions

In this section, we assume that the reader is familiar with partial differentiation: that is, if a differentiable function $u(x, y)$ depends on the variables x and y then *partial derivatives* with respect to both x and y may be calculated. Readers unfamiliar with partial differential equations may wish to skip this section or consult a suitable text on mathematical methods such as Kreyszig [2].

In the previous section, we designed a library for calculating the finite difference solution of linear, constant coefficient, second order, boundary value ordinary differential equations. We will now explain how a library may be developed for the finite difference solution of Poisson's equation in two spatial dimensions on a rectangular domain, with Dirichlet boundary conditions, that is, equations of the form

$$\frac{\partial^2 u}{\partial x^2} + \frac{\partial^2 u}{\partial y^2} = f(x, y), \qquad X_0 < x < X_1, Y_0 < y < Y_1,$$

where X_0, X_1, Y_0, Y_1 are specified constants, $f(x, y)$ is a specified function, and boundary conditions for u are given at each point on the boundary of the rectangular domain specified.

12.3.1 Model Problem for Two Dimensions

As with ordinary differential equations earlier in this chapter, we will use a model problem to demonstrate the implementation of the finite difference method. The model problem that we will use is

$$\frac{\partial^2 u}{\partial x^2} + \frac{\partial^2 u}{\partial y^2} = -4(1 - x^2 - y^2)e^{-(x^2 + y^2)}, \quad 0 < x < 1, 0 < y < 2, \tag{12.11}$$

$$u = e^{-y^2}, \quad x = 0, \ 0 < y < 2, \tag{12.12}$$

$$u = e^{-(1+y^2)}, \quad x = 1, \ 0 < y < 2, \tag{12.13}$$

$$u = e^{-x^2}, \quad 0 < x < 1, y = 0, \tag{12.14}$$

$$u = e^{-(4+x^2)}, \quad 0 < x < 1, \ y = 2. \tag{12.15}$$

This model problem has solution

$$u = e^{-(x^2+y^2)}.$$

12.3.2 Finite Difference Methods for Boundary Value Problems in Two Dimensions

To define the finite differences that approximate the partial derivatives of a function in two dimensions, we first need to define a finite difference grid. We have already stated that we are assuming that the function u that is to be determined satisfies a partial differential equation defined on the region $X_0 \leq x \leq X_1, Y_0 \leq y \leq Y_1$. We now suppose that there are points $x_i, i = 1, 2, \ldots, M$ and $y_j, j = 1, 2, \ldots, N$ such that

$$x_1 = X_0,$$
$$x_1 < x_2 < x_3 < \cdots < x_M,$$
$$x_M = X_1,$$
$$y_1 = Y_0,$$
$$y_1 < y_2 < y_3 < \cdots < y_N,$$
$$y_N = Y_1.$$

The nodes of the finite difference grid are then the points $(x_i, y_j), i = 1, 2, \ldots, M$ $j = 1, 2, \ldots, N$. The boundary nodes are the nodes where $x = X_0, x = X_1, y = Y_0$ or $y = Y_1$. All other nodes are interior nodes. An example mesh on the square $0 <$ $x < 1, 0 < y < 2$ is shown in Fig. 12.2, where the filled circles denote the boundary nodes, and the open circles denote the interior nodes.

Numbering of the nodes for a finite difference grid is slightly more complicated in two dimensions than it was in one dimension. For the finite difference grid in one dimension all nodes could be numbered consecutively, allowing the finite difference approximations to be written down in an intuitive way. To write down finite difference approximations in two dimensions, we will adopt the "compass point" notation shown in Fig. 12.3. The node immediately above node i in the computational mesh is denoted by i, N, where "N" corresponds to north. The other nodes that are adjacent to node i are the *east*, *south* and *west* nodes, denoted by "i, E", "i, S" and "i, W" respectively

Provided i is an interior node, the four adjacent nodes shown in Fig. 12.3 all exist Finite differences to the derivatives that appear in Poisson's equation are given below

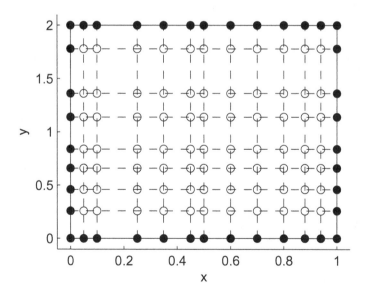

Fig. 12.2 A suitable finite difference grid in two dimensions. Boundary nodes are denoted by a *filled circle*, interior nodes by a *hollow circle*

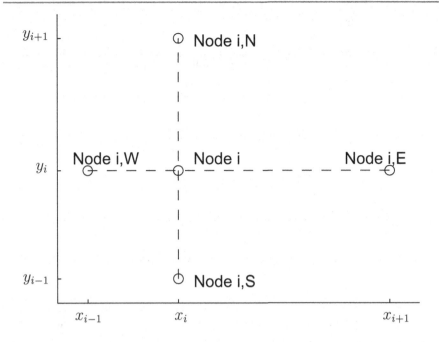

Fig. 12.3 Node i and points used to calculate finite difference approximations in two dimensions

$$\frac{\partial^2 u}{\partial x^2} \approx \frac{2}{x_{i,E} - x_{i,W}} \left(\frac{u_{i,E} - u_i}{x_{i,E} - x_i} - \frac{u_i - u_{i,W}}{x_i - x_{i,W}} \right), \qquad (12.16)$$

$$\frac{\partial^2 u}{\partial y^2} \approx \frac{2}{y_{i,N} - y_{i,S}} \left(\frac{u_{i,N} - u_i}{y_{i,N} - y_i} - \frac{u_i - u_{i,S}}{y_i - y_{i,S}} \right). \qquad (12.17)$$

We will now explain how these finite difference approximations may be used to set up a linear system to calculate the numerical solution of Poisson's equation.

12.3.3 Setting Up the Linear System for the Model Problem

We will now apply the theory developed in Sect. 12.3.2 to the model problem described in Sect. 12.3.1. Using the finite difference grid described in Sect. 12.3.2, we have M nodes in the x-direction, and N nodes in the y-direction: that is, a total of $M \times N$ nodes. Each of these nodes has an unknown value of u, and so our linear system comprises $M \times N$ equations, with each equation being associated with one node of the mesh.

At interior nodes we may substitute the finite difference approximations given in Eqs. (12.16) and (12.17). Substituting these approximation into Eq. (12.11) and rearranging yields

$$\alpha_i u_i + \alpha_{i,N} u_{i,N} + \alpha_{i,E} u_{i,E} + \alpha_{i,S} u_{i,S} + \alpha_{i,W} u_{i,W} = b_i, \qquad (12.18)$$

where

$$\alpha_i = -\frac{2}{(x_{i,E} - x_i)(x_i - x_{i,W})} - \frac{2}{(y_{i,N} - y_i)(y_i - y_{i,S})},$$

$$\alpha_{i,N} = \frac{2}{(y_{i,N} - y_{i,S})(y_{i,N} - y_i)},$$

$$\alpha_{i,E} = \frac{2}{(x_{i,E} - x_{i,W})(x_{i,E} - x_i)},$$

$$\alpha_{i,S} = \frac{2}{(y_{i,N} - y_{i,S})(y_i - y_{i,S})},$$

$$\alpha_{i,W} = \frac{2}{(x_{i,E} - x_{i,W})(x_i - x_{i,W})},$$

$$b_i = -4(1 - x_i^2 - y_i^2)e^{-(x_i^2 + y_i^2)}.$$

The value of u at each boundary node is given by the appropriate equation from Eqs. (12.12)–(12.15). This may be incorporated into the linear system by the equation

$$u_i = b_i, \tag{12.19}$$

where i is a boundary node, and b_i is the value that u takes at that node.

Equations (12.18) and (12.19) fully define the linear system. We may now use the functionality of the classes of vectors, matrices and linear systems developed in Chap. 10 to calculate the value of the finite difference approximation to u at each node.

12.3.4 Developing the Classes Required

We give only minimal guidance on developing the classes required for calculating a numerical solution of Poisson's equation. Designing and implementing these classes is left as an exercise (Exercise 12.4). Our suggestions are given below.

- Creating an instance of the class FiniteDifferenceGrid should require the use of a constructor that specifies the number of nodes in the x direction and the number of nodes in the y direction. The grid should consist of a vector of boundary nodes that are all instances of the class BoundaryNode (discussed below) and a vector of interior nodes that are all members of the class InteriorNode (also discussed below). Each of the nodes in the mesh should have a global numbering that will refer to the row number of the matrix that will correspond to the unknown value of u at that node, u_i.
- An instance of the class BoundaryNode will have an integer representing the global numbering, and a double precision floating point variable that represents the value of u at that node from the boundary conditions.

- An instance of the class `InteriorNode` will have an integer representing the global numbering, and the global numbers of the north node, east node, south node and west node: see Fig. 12.3 for a definition of these nodes.

The classes described above, together with a class for encapsulating the partial differential equation that is similar to `SecondOrderOde` in Sect. 12.2, should enable code to be written to calculate the numerical solution of Poisson's equation.

12.4   Tips: Using Well-Written Libraries

In Chap. 10 we developed a linear system class that was based on classes of vectors and matrices. These classes allowed us to perform various linear algebra operations. In this chapter, we utilised these classes to allow us to develop libraries for calculating the numerical solution of boundary value ordinary differential equations.

Although the classes developed in Chap. 10 do have sufficient functionality for the purpose of this chapter, we would recommend that a reader who requires a linear algebra library should consider using one of the many high quality, open-source libraries that are available. (Indeed, in Sect. 1.1.2, we gave the fact that there is a wealth of numerical libraries for scientific computing as one of the reasons for learning C++.) Libraries for linear algebra usually include significantly more functionality than that developed here including, for example: sparse matrices; a wide variety of iterative linear solvers; a wide variety of preconditioners; interfaces with other packages; and support for parallelisation. Indeed, as linear algebra is such a fundamental topic at the core of scientific computing, it is unlikely that any functionality required will not be included in a widely used library. Furthermore, such libraries have the advantage of being well-tested, optimised code and can, as such, be treated as a black box.

One open-source library that is of particular use is the Portable Extensible Toolkit for Scientific Computing (PETSc, pronounced "pet see") which is available for download from https://www.mcs.anl.gov/petsc/. Libraries such as PETSc include an extremely large amount of functionality for systems of both linear and nonlinear equations, with support for parallel implementation on distributed memory architectures through the MPI library.

We conclude this section by reminding the reader of our the remarks in Sect. 1.1.4. We explained in that section that this book focuses on aspects of the C++ programming language that are commonly needed when writing software for scientific computing applications. As such, we haven't touched on the functionality of the language that is rarely required in this field. Should readers wish to develop their C++ skills to use more advanced features we have given a list of suitable references in the Further Reading at the end of this book [5–10].

12.5   Exercises

12.1 Develop the classes described in Sect. 12.2 for second order, constant coefficient, linear boundary value ordinary differential equations. Test these libraries using

the model problems described in Sect. 12.1.1. The code in Listing 12.6 which pro
duces output files that can be readily plotted may be used as a framework. Example
solutions for this problem are given in Sect. C.2: these files, together with the header
files given in this chapter, may be downloaded from http://www.springer.com/book.
9783319731315.

Make sure that the `BvpOde` method `WriteSolutionFile` does not attempt
to write a file if `mFilename` is uninitialised or set to an empty string. (This may be
achieved by setting `mFilename` to a safe value in the constructor.)

12.2 Extend the library developed in Exercise 12.1 so that the user may specify a
nonuniform finite difference grid. Allow this to be done through a method `SetGrid`
of the class `FiniteDifferenceGrid` that allows a mesh to be specified as a
vector of ordered nodes. Ensure that the boundary nodes have the same value as
`mXmin` and `mXmax` in the class `SecondOrderOde`.

12.3 Some programmers may feel that the constructor given in Listing 12.1 is inad-
equate. They may argue that it would be easy to incorrectly assign one of the coefficients of the equation. One way around this would be to force the user to use a default
constructor. Additional class members, such as a Boolean variable `mCoeffOf-
UxxIsSet` could be deployed. The default constructor would be overridden so that
these variables were set to `false` when the constructor was called. A method called
`SetCoefficientOfUxx` would then be written, which would have as input the
coefficient of d^2u/dx^2. This method would assign the coefficient correctly and set
the Boolean variable `mCoeffOfUxxIsSet` to `true`. Before the methods that calculate the numerical solution are called a check would be carried out to ensure that
all required data has been assigned. Design, and implement, classes to specify the
differential equation in this way.

12.4 If you understand the theory for finite difference methods for Poisson's equation given in Sect. 12.3.2, develop a library for solving such equations. Test this
library using the model problem described in Sect. 12.3.1.

12.5 Exercise 12.1 asks you to develop the classes described in Sect. 12.2 and to
test these libraries using the model problems described in Sect. 12.1.1. For this
purpose Listing 12.6 gives a program `Driver.cpp`. This way of "testing" is not
ideal because it relies on the manual step of checking that the data in the output files
matches the expected solution.

Automate the process of testing the classes described in Sect. 12.2 by rewriting the
testing functionality within a testing framework such as `CxxTest`. For each model
problem you should produce a testing function which runs the problem, reads the
output file back into a suitable data structure, and tests that the solution is correct:
that is, the solution is close to the analytic form given in Sect. 12.1.1. Think about
what the expected error might be for this numerical scheme.

An example solution to this problem is given in Listing C.9 in Sect. C.2.

Linear Algebra

This appendix summarises the linear algebra that underpins the classes of vectors and matrices developed in this book. We present little more than the algorithms used: a reader interested in a deeper understanding of this theory should consult a textbook such as one of those listed in the Further Reading section at the end of this book.

A.1 Vectors and Matrices

For the purpose of this book, a vector is a one-dimensional array and a matrix is a two-dimensional array: it is—of course—possible to work only with matrices, with vectors having either only one column or only one row. For consistency with the classes of vectors and matrices developed, we treat vectors and matrices as separate entities in this discussion.

In this Appendix, we use mathematical rather than C++ notation for vectors and matrices. We will use italics to denote a scalar. Vectors will be denoted by lower case bold font letters. Individual entries of a vector will be denoted by italics indexed by subscripts. For example, $\mathbf{v}$ represents a vector, and the entry of $\mathbf{v}$ with index i is denoted by v_i. For consistency with C++ coding, we index the vectors and matrices in this Appendix so that the indices begin from 0. We assume that all vectors are column vectors: that is, a vector $\mathbf{v}$ of length N is the vector

$$\mathbf{v} = \begin{pmatrix} v_0 \\ v_1 \\ \vdots \\ v_{N-1} \end{pmatrix}.$$

If a row vector is required, it is denoted using the transpose superscript, that is, $\mathbf{v}^\top$. Matrices will be denoted by upper case bold font letters, with italics indexed by

© Springer International Publishing AG, part of Springer Nature 2017
J. Pitt-Francis and J. Whiteley, *Guide to Scientific Computing in C++*, Undergraduate Topics in Computer Science,
https://doi.org/10.1007/978-3-319-73132-2

subscripts used to denote the entries of the matrix. The first index corresponds to the row number and the second index corresponds to the column number. Using this notation, if $\mathbf{A}$ is a matrix, then the entry of $\mathbf{A}$ that appears in the row with index i and the column with index j is denoted by A_{ij}. Where required for clarity, we will separate the indices by a comma, for example $A_{i+1,j-1}$.

A square matrix of size N has both N rows and N columns. The identity matrix of size N is a square matrix, denoted by $\mathbf{I}^{(N)}$, with entries given by

$$I_{ij}^{(N)} = \begin{cases} 1, & i = j, \\ 0, & i \neq j. \end{cases}$$

A.1.1 Operations Between Vectors and Matrices

Linear combinations of vectors. Suppose $\mathbf{w} = \alpha\mathbf{u} + \beta\mathbf{v}$, where $\mathbf{u}, \mathbf{v}, \mathbf{w}$ are all vectors of length N, and α, β are scalars. The entries of $\mathbf{w}$ are given by

$$w_i = \alpha u_i + \beta v_i, \quad i = 0, 1, \ldots, N - 1.$$

Linear combinations of matrices. Suppose $\mathbf{C} = \alpha\mathbf{A} + \beta\mathbf{B}$, where $\mathbf{A}, \mathbf{B}, \mathbf{C}$ are all matrices with M rows and N columns, and α, β are scalars. The entries of $\mathbf{C}$ are given by

$$C_{ij} = \alpha A_{ij} + \beta B_{ij}, \quad i = 0, 1, \ldots, M - 1, \quad j = 0, 1, \ldots, N - 1.$$

Multiplication of a matrix by a vector. Suppose $\mathbf{A}$ is a matrix with M rows and N columns, and $\mathbf{u}$ is a vector of length N. If $\mathbf{v} = \mathbf{A}\mathbf{u}$, then $\mathbf{v}$ is a vector of length M with entries given by

$$v_i = \sum_{j=0}^{N-1} A_{ij} u_j, \quad i = 0, 1, \ldots, M - 1.$$

Similarly, if $\mathbf{s}$ is a vector of length M and $\mathbf{t}^\top = \mathbf{s}^\top \mathbf{A}$, then $\mathbf{t}$ is a vector of length M with entries given by

$$t_j = \sum_{i=0}^{M-1} s_i A_{ij}, \quad j = 0, 1, \ldots, N - 1.$$

Multiplication of a matrix by a matrix. Suppose $\mathbf{A}$ is a matrix with L rows and M columns, and $\mathbf{B}$ is a matrix with M rows and N columns. If the matrix $\mathbf{C}$ satisfies $\mathbf{C} = \mathbf{A}\mathbf{B}$, then $\mathbf{C}$ has L rows and N columns, and has entries given by

$$C_{ij} = \sum_{k=0}^{M-1} A_{ik} B_{kj}, \quad i = 0, 1, \ldots, L - 1, \quad j = 0, 1, \ldots, N - 1.$$

The transpose of a matrix. Suppose $\mathbf{A}$ is a matrix with M rows and N columns. If the matrix $\mathbf{B}$ satisfies $\mathbf{B} = \mathbf{A}^\top$, then $\mathbf{B}$ has N rows and M columns with entries given by

$$B_{ij} = A_{ji}, \quad i = 0, 1, \ldots, N - 1, \quad j = 0, 1, \ldots, M - 1.$$

A matrix $\mathbf{A}$ is said to be symmetric if $\mathbf{A} = \mathbf{A}^\top$.

A.1.2 The Scalar Product of Two Vectors

Suppose $\mathbf{v}$ and $\mathbf{w}$ are both vectors of length N. The *scalar product* between $\mathbf{v}$ and $\mathbf{w}$, denoted by $\mathbf{v} \cdot \mathbf{w}$, is given by

$$\mathbf{v} \cdot \mathbf{w} = \sum_{i=0}^{N-1} v_i w_i. \tag{A.1}$$

A.1.3 The Determinant and the Inverse of a Matrix

The simplest way to specify the determinant of a square matrix of general size is to use recursion.[1] Suppose $\mathbf{A}$ is a square matrix of size N. The determinant of $\mathbf{A}$, denoted by $\det(\mathbf{A})$, may be written

$$\det(\mathbf{A}) = A_{00} \det(\hat{\mathbf{A}}^{(00)}) - A_{01} \det(\hat{\mathbf{A}}^{(01)}) + A_{02} \det(\hat{\mathbf{A}}^{(02)}) - A_{03} \det(\hat{\mathbf{A}}^{(03)}) + \cdots$$
$$+ (-1)^{N-1} A_{0,N-1} \det(\hat{\mathbf{A}}^{(0,N-1)}),$$

where the square matrix $\hat{\mathbf{A}}^{(ij)}$, of size $N - 1$, is the matrix $\mathbf{A}$ with row i and column j removed. This definition allows us to express the determinant of a square matrix of size N as a sum of determinants of square matrices of size $N - 1$. This process may be repeated recursively until the determinant is expressed as a sum of determinants of square matrices of size 1. To complete this definition, we need to define the determinant of a square matrix of size 1: under these conditions $\det(\mathbf{A}) = A_{00}$. We leave it to the reader to verify that this definition is consistent with the commonly used expressions for the determinant of matrices of sizes 2 and 3.

---

[1] This recursion may be mapped directly into recursive functions (discussed in Sect. 5.8) when programming. However, it is generally more efficient to hard-code commonly used determinants for small matrices such as 2×2 and 3×3.

If the determinant of a square matrix $\mathbf{A}$ of size N is nonzero, then $\mathbf{A}$ is said to be *invertible*: a unique inverse matrix—denoted by $\mathbf{A}^{-1}$—exists, and satisfies

$$\mathbf{A}^{-1}\mathbf{A} = \mathbf{A}\mathbf{A}^{-1} = \mathbf{I}^{(N)}.$$

For the square matrix $\mathbf{A}$ of size 2 given by

$$\mathbf{A} = \begin{pmatrix} a & b \\ c & d \end{pmatrix},$$

then provided the determinant, given by $ad - bc$ is nonzero, $\mathbf{A}^{-1}$ exists and is given by

$$\mathbf{A}^{-1} = \frac{1}{ad - bc} \begin{pmatrix} d & -b \\ -c & a \end{pmatrix}.$$

A.1.4 Eigenvalues and Eigenvectors of a Matrix

Suppose $\mathbf{A}$ is a square matrix of size N. The scalar λ is said to be an eigenvalue of $\mathbf{A}$ if

$$\det(\mathbf{A} - \lambda \mathbf{I}^{(N)}) = 0.$$

If λ is an eigenvalue of A then a family of nonzero vectors[2] $\mathbf{v}$ that satisfy $\mathbf{A}\mathbf{v} = \lambda \mathbf{v}$ exists: each $\mathbf{v}$ in this family is then said to be an *eigenvector* corresponding to the eigenvalue λ.

A.1.5 Vector and Matrix Norms

Suppose $\mathbf{v}$ is a vector of length N. The p-norm of $\mathbf{v}$, denoted by $\|\mathbf{v}\|_p$, is given by

$$\|\mathbf{v}\|_p = \left(\sum_{i=0}^{N-1} |v_i|^p \right)^{1/p}. \tag{A.2}$$

---

[2]A vector $\mathbf{v}$ satisfies $\mathbf{v} = \mathbf{0}$ if, and only if, all entries of this vector take the value 0: $\mathbf{v}$ is then said to be a *zero vector*. If not, $\mathbf{v}$ is said to be a *nonzero vector*.

Taking the limit as $p \to \infty$, this definition yields

$$\|\mathbf{v}\|_\infty = \max_{i=0}^{N-1} |v_i| .$$

Of most use is the 2-norm: this is known as the Euclidean norm, and corresponds to the length of the line that represents a vector in two or three dimensions. Using Eq. (A.1), and Eq. (A.2) with $p = 2$, we see that we may write the 2-norm as

$$\|\mathbf{v}\|_2 = \sqrt{\sum_{i=0}^{N-1} v_i^2} = \sqrt{\mathbf{v} \cdot \mathbf{v}}.$$

The p-norm of a matrix $\mathbf{A}$, denoted by $\|\mathbf{A}\|_p$, is given (in terms of the vector p-norm) by

$$\|\mathbf{A}\|_p = \max_{\mathbf{v} \neq 0} \frac{\|\mathbf{A}\mathbf{v}\|_p}{\|\mathbf{v}\|_p}.$$

In common with vector norms, the most commonly used norm is the 2-norm. It can be shown that the eigenvalues of the matrix $\mathbf{A}^\top \mathbf{A}$ are all real and nonnegative. Let λ be the largest of these eigenvalues. Then $\|\mathbf{A}\|_2 = \sqrt{\lambda}$.

A.2 Systems of Linear Equations

Many algorithms in scientific computing require the solution of linear systems of the form $\mathbf{A}\mathbf{x} = \mathbf{b}$, where: (i) $\mathbf{A}$ is a square, invertible matrix of size N; (ii) the vectors $\mathbf{x}$, $\mathbf{b}$ are both of length N; (iii) $\mathbf{A}$, $\mathbf{b}$ are known; and (iv) $\mathbf{x}$ is to be calculated. Clearly $\mathbf{x}$ satisfies $\mathbf{x} = \mathbf{A}^{-1}\mathbf{b}$. However, calculating $\mathbf{A}^{-1}$ is extremely computationally expensive for large N and this approach is rarely used to solve systems of linear equations. Instead a plethora of techniques are available: we list three relatively simple methods below.

A.2.1 Gaussian Elimination

Readers may remember being taught how to solve two simultaneous linear equations for unknown values of x and y at school. When using this technique, the first step is to eliminate one of the variables resulting in a single linear equation for a single variable that can easily be solved. The value of this variable is then substituted back into one of the original equations to allow the value of the other variable to be calculated. Gaussian elimination is a systematic extension of this technique when solving a system of N linear equations for N unknowns. There are two versions of Gaussian elimination: with or without pivoting. We now describe both of these versions.

A.2.1.1 Gaussian Elimination Without Pivoting

The original system of equations may be written

$$
\begin{pmatrix}
A_{00} & A_{01} & A_{02} & \cdots & A_{0,N-1} \\
A_{10} & A_{11} & A_{12} & \cdots & A_{1,N-1} \\
A_{20} & A_{21} & A_{22} & \cdots & A_{2,N-1} \\
\vdots & \vdots & \vdots & \ddots & \vdots \\
A_{N-1,0} & A_{N-1,1} & A_{N-1,2} & \cdots & A_{N-1,N-1}
\end{pmatrix}
\begin{pmatrix}
x_0 \\ x_1 \\ x_2 \\ \vdots \\ x_{N-1}
\end{pmatrix}
=
\begin{pmatrix}
b_0 \\ b_1 \\ b_2 \\ \vdots \\ b_{N-1}
\end{pmatrix}.
$$

Let us first assume that $A_{00} \neq 0$. This is a very restrictive assumption: in Sect A.2.1.3 we introduce *pivoting*, which allows us to deal with the case $A_{00} = 0$. The assumption $A_{00} \neq 0$ allows us to eliminate x_0 from all but the first equation: this is achieved by subtracting a suitable multiple of the first equation, and results in the following system:

$$
\begin{pmatrix}
A_{00} & A_{01} & A_{02} & \cdots & A_{0,N-1} \\
0 & A_{11}^{(1)} & A_{12}^{(1)} & \cdots & A_{1,N-1}^{(1)} \\
0 & A_{21}^{(1)} & A_{22}^{(1)} & \cdots & A_{2,N-1}^{(1)} \\
\vdots & \vdots & \vdots & \ddots & \vdots \\
0 & A_{N-1,1}^{(1)} & A_{N-1,2}^{(1)} & \cdots & A_{N-1,N-1}^{(1)}
\end{pmatrix}
\begin{pmatrix}
x_0 \\ x_1 \\ x_2 \\ \vdots \\ x_{N-1}
\end{pmatrix}
=
\begin{pmatrix}
b_0 \\ b_1^{(1)} \\ b_2^{(1)} \\ \vdots \\ b_{N-1}^{(1)}
\end{pmatrix},
$$

where:

$$
M_{i0} = A_{i0}/A_{00}, \quad i = 1, 2, \ldots, N-1, \text{ using the assumption that } A_{00} \neq 0,
$$
$$
A_{ij}^{(1)} = A_{ij} - M_{i0}A_{0j}, \quad i, j = 1, 2, \ldots, N-1,
$$
$$
b_i^{(1)} = b_i - M_{i0}b_0, \quad i = 1, 2, \ldots, N-1.
$$

Assuming now that $A_{11}^{(1)} \neq 0$, we may repeat this process to eliminate x_1 from all but the first two equations:

$$
\begin{pmatrix}
A_{00} & A_{01} & A_{02} & \cdots & A_{0,N-1} \\
0 & A_{11}^{(1)} & A_{12}^{(1)} & \cdots & A_{1,N-1}^{(1)} \\
0 & 0 & A_{22}^{(2)} & \cdots & A_{2,N-1}^{(2)} \\
\vdots & \vdots & \vdots & \ddots & \vdots \\
0 & 0 & A_{N-1,2}^{(2)} & \cdots & A_{N-1,N-1}^{(2)}
\end{pmatrix}
\begin{pmatrix}
x_0 \\ x_1 \\ x_2 \\ \vdots \\ x_{N-1}
\end{pmatrix}
=
\begin{pmatrix}
b_0 \\ b_1^{(1)} \\ b_2^{(2)} \\ \vdots \\ b_{N-1}^{(2)}
\end{pmatrix},
$$

where:

$$
M_{i1} = A_{i1}^{(1)}/A_{11}^{(1)}, \quad i = 2, 3, \ldots, N-1, \text{ using the assumption that } A_{11}^{(1)} \neq 0,
$$
$$
A_{ij}^{(2)} = A_{ij}^{(1)} - M_{i1}A_{1j}^{(1)}, \quad i, j = 2, 3, \ldots, N-1,
$$
$$
b_i^{(2)} = b_i^{(1)} - M_{i1}b_1^{(1)}, \quad i = 2, 3, \ldots, N-1.
$$

Providing that at all steps we have $A_{kk}^{(k)} \neq 0, k = 0, 1, \ldots, N - 1$, we may continue in this fashion until we have generated an upper triangular matrix $\mathbf{A}^{(N-1)}$:

$$
\mathbf{A}^{(N-1)}\mathbf{x} = \begin{pmatrix} A_{00} & A_{01} & A_{02} & \cdots & A_{0,N-1} \\ 0 & A_{11}^{(1)} & A_{12}^{(1)} & \cdots & A_{1,N-1}^{(1)} \\ 0 & 0 & A_{22}^{(2)} & \cdots & A_{2,N-1}^{(2)} \\ \vdots & \vdots & \vdots & \ddots & \vdots \\ 0 & 0 & 0 & \cdots & A_{N-1,N-1}^{(N-1)} \end{pmatrix} \begin{pmatrix} x_0 \\ x_1 \\ x_2 \\ \vdots \\ x_{N-1} \end{pmatrix} = \begin{pmatrix} b_0 \\ b_1^{(1)} \\ b_2^{(2)} \\ \vdots \\ b_{N-1}^{(N-1)} \end{pmatrix} = \mathbf{b}^{(N-1)}.
$$

Solving this upper triangular system is a straightforward task: we start with the last equation in this system and work our way backwards. The first two steps in this procedure are

$$
x_{N-1} = b_{N-1}^{(N-1)}/A_{N-1,N-1}^{(N-1)},
$$

$$
x_{N-2} = \frac{1}{A_{N-2,N-2}^{(N-2)}} \left(b_{N-2}^{(N-2)} - A_{N-2,N-1}^{(N-2)} x_{N-1} \right).
$$

A general formula exists for calculating $x_k, k = 0, 1, 2, \ldots, N - 1$. Assuming that we have already calculated $x_{k+1}, x_{k+2}, \ldots, x_{N-1}$, we may calculate x_k by

$$
x_k = \frac{1}{A_{k,k}^{(k)}} \left(b_k^{(k)} - \sum_{i=k+1}^{N-1} A_{k,i}^{(k)} x_i \right). \tag{A.3}
$$

This completes the description of the Gaussian elimination algorithm without pivoting. A very important point to note is that there is no need to store all the matrices generated during this algorithm: only the most recently generated version is required, and all earlier matrices may be discarded.

A.2.1.2 LU Decomposition

The Gaussian elimination process described above may be used to factorise $\mathbf{A}$ as the product of a lower triangular matrix $\mathbf{L}$ and an upper triangular matrix $\mathbf{U}$, that is, $\mathbf{A} = \mathbf{LU}$. Defining the matrices $\mathbf{M}_0, \mathbf{M}_1, \ldots$ by

$$
\mathbf{M}_0 = \begin{pmatrix} 1 & 0 & 0 & \cdots & 0 \\ -M_{10} & 1 & 0 & \cdots & 0 \\ -M_{20} & 0 & 1 & \cdots & 0 \\ \vdots & \vdots & \vdots & \ddots & \vdots \\ -M_{N-1,0} & 0 & 0 & \cdots & 1 \end{pmatrix},
$$

$$
\mathbf{M}_1 = \begin{pmatrix} 1 & 0 & 0 & \cdots & 0 \\ 0 & 1 & 0 & \cdots & 0 \\ 0 & -M_{21} & 1 & \cdots & 0 \\ \vdots & \vdots & \vdots & \ddots & \vdots \\ 0 & -M_{N-1,1} & 0 & \cdots & 1 \end{pmatrix}, \ldots,
$$

we may write

$$\mathbf{A}^{(N-1)} = \mathbf{M}_{N-1}\mathbf{M}_{N-2}\cdots\mathbf{M}_1\mathbf{M}_0\mathbf{A},$$

or, equivalently,

$$\mathbf{A} = \mathbf{M}_0^{-1}\mathbf{M}_1^{-1}\cdots\mathbf{M}_{N-2}^{-1}\mathbf{M}_{N-1}^{-1}\mathbf{A}^{(N-1)}.$$

We first note that the inverses of the matrices $\mathbf{M}_0, \mathbf{M}_1, \ldots$, are simply

$$\mathbf{M}_0^{-1} = \begin{pmatrix} 1 & 0 & 0 & \ldots & 0 \\ M_{10} & 1 & 0 & \ldots & 0 \\ M_{20} & 0 & 1 & \ldots & 0 \\ \vdots & \vdots & \vdots & \ddots & \vdots \\ M_{N-1,0} & 0 & 0 & \ldots & 1 \end{pmatrix},$$

$$\mathbf{M}_1^{-1} = \begin{pmatrix} 1 & 0 & 0 & \ldots & 0 \\ 0 & 1 & 0 & \ldots & 0 \\ 0 & M_{21} & 1 & \ldots & 0 \\ \vdots & \vdots & \vdots & \ddots & \vdots \\ 0 & M_{N-1,1} & 0 & \ldots & 1 \end{pmatrix}, \ldots.$$

These matrices are all lower triangular. It is trivial to prove that the product of lower triangular matrices is also lower triangular. Writing

$$\mathbf{L} = \mathbf{M}_0^{-1}\mathbf{M}_1^{-1}\cdots\mathbf{M}_{N-2}^{-1}\mathbf{M}_{N-1}^{-1},$$
$$\mathbf{U} = \mathbf{A}^{(N-1)},$$

we see that we have $\mathbf{A} = \mathbf{LU}$ with $\mathbf{L}$ a lower triangular matrix and $\mathbf{U}$ an upper triangular matrix. An explicit representation of $\mathbf{L}$ exists: direct calculation may be used to verify that

$$\mathbf{L} = \begin{pmatrix} 1 & 0 & 0 & \ldots 0 \\ M_{10} & 1 & 0 & \ldots 0 \\ M_{20} & M_{21} & 1 & \ldots 0 \\ \vdots & \vdots & \vdots & \ddots \vdots \\ M_{N-1,0} & M_{N-1,1} & M_{N-1,2} & \ldots 1 \end{pmatrix}.$$

A.2.1.3 Gaussian Elimination with Pivoting

The Gaussian elimination technique described above required that $A_{kk}^{(k)} \neq 0$ at each step. Clearly this algorithm would fail for a nonsingular matrix such as

$$\mathbf{A} = \begin{pmatrix} 1 & 1 & 1 \\ 1 & 1 & 2 \\ 0 & 5 & 1 \end{pmatrix},$$

where

$$\mathbf{A}^{(1)} = \begin{pmatrix} 1 & 1 & 1 \\ 0 & 0 & 1 \\ 0 & 5 & 1 \end{pmatrix},$$

and so $A_{11}^{(1)} = 0$, violating one of the assumptions made in Sect. A.2.1.1. We can, however, proceed further: in this case we would simply interchange the last two rows of both $\mathbf{A}^{(1)}$ and $\mathbf{b}^{(1)}$. This is known as *pivoting*.

Even if $|A_{kk}^{(k)}|$ is not zero it may be advisable to use pivoting. In Eq. (A.3) we see that calculating the value of x_k requires us to divide by $A_{kk}^{(k)}$. If $|A_{kk}^{(k)}|$ is small then the division by a small number may introduce numerical errors in the calculation of x_k. To avoid both of these problems, we recommend pivoting at each step: when constructing $\mathbf{A}^{(k)}$, find the row n with the largest absolute value of $A_{nk}^{(k)}$, $n = k, k+1, \ldots, N-1$, and then interchange row k and row n. It is relatively simple to include this in our Gaussian elimination algorithm: at step k we are working with the linear system

$$\mathbf{A}^{(k)}\mathbf{x} = \mathbf{b}^{(k)}.$$

To interchange rows k and n in this system of equations, we simply multiply both sides of this equation by the matrix $\mathbf{P}^{(kn)}$:

$$\mathbf{P}^{(kn)}\mathbf{A}^k\mathbf{x} = \mathbf{P}^{(kn)}\mathbf{b}^k,$$

where $\mathbf{P}^{(kn)}$ is a square matrix of size N with entries given by

$$\mathbf{P}_{ij}^{(kn)} = \begin{cases} 1, & i = j, \ i, j \neq k, \ i, j \neq n, \\ 1, & i = k, \ j = n, \\ 1, & i = n, \ j = k, \\ 0, & \text{otherwise.} \end{cases}$$

For example, if we wanted to interchange the row with index 2 and the row with index 4 in a square matrix of size 5, then the matrix $\mathbf{P}^{(24)}$ would be given by

$$\mathbf{P}^{(24)} = \begin{pmatrix} 1 & 0 & 0 & 0 & 0 \\ 0 & 1 & 0 & 0 & 0 \\ 0 & 0 & 0 & 0 & 1 \\ 0 & 0 & 0 & 1 & 0 \\ 0 & 0 & 1 & 0 & 0 \end{pmatrix}.$$

The key point to note when modifying the LU-factorisation algorithm described in Sect. A.2.1.1 to take account of pivoting is that Gaussian elimination with pivoting would give exactly the same results if all the rows were interchanged first, and then Gaussian elimination with no pivoting were carried out. Denoting the product of all the matrices representing row interchanges by $\mathbf{P}$, we see that the LU-decomposition algorithm now reduces to a factorisation of the matrix $\mathbf{PA}$: that is, forming

$$\mathbf{LU} = \mathbf{PA}.$$

A.2.2 The Thomas Algorithm

The Thomas algorithm may be used for matrices with a specific structure. Suppose
our matrix $\mathbf{A}$ has structure

$$
\mathbf{A} = \begin{pmatrix}
1 & 0 & 0 & 0 & \cdots & 0 & 0 & 0 \\
-p_1 & q_1 & -r_1 & 0 & \cdots & 0 & 0 & 0 \\
0 & -p_2 & q_2 & -r_2 & \cdots & 0 & 0 & 0 \\
\vdots & \vdots & \vdots & \vdots & \ddots & \vdots & \vdots & \vdots \\
0 & 0 & 0 & 0 & \cdots & -p_{N-2} & q_{N-2} & -r_{N-2} \\
0 & 0 & 0 & 0 & \cdots & 0 & 0 & 1
\end{pmatrix},
$$

where the entries of $\mathbf{A}$ satisfy

$$
p_i > 0, \qquad q_i > 0, \qquad r_i > 0, \qquad q_i > p_i + r_i, \qquad i = 1, 2, \ldots, N - 2.
$$

This condition is satisfied, for example, for an implicit finite difference discretisation
of the heat equation in one spatial dimension with Dirichlet boundary conditions at
both ends of the spatial domain. Defining

$$
e_0 = 0, \qquad f_0 = b_0,
$$
$$
e_i = \frac{r_i}{q_i - p_i e_{i-1}}, \qquad f_i = \frac{b_i + p_i f_{i-1}}{q_i - p_i e_{i-1}}, \qquad i = 1, 2, \ldots, N - 2,
$$

then the linear system may be solved using the explicit recurrence relation

$$
x_{N-1} = b_{N-1},
$$
$$
x_i = \frac{r_i}{q_i - p_i e_{i-1}} x_{i+1} + \frac{b_i + p_i f_{i-1}}{q_i - p_i e_{i-1}}, \qquad i = N - 2, N - 3, \ldots, 1,
$$
$$
x_0 = b_0.
$$

A.2.3 The Conjugate Gradient Method

The matrices arising in many scientific computing applications—for example,
finite element, finite difference and finite volume discretisations of partial differential
equations—often have a large number of rows and columns, but very few nonzero
elements in each row of the matrix. Such matrices are termed *sparse matrices*.

It is often the case that storing every element of a sparse matrix would exceed the
memory limitations of a computational architecture, but storing only the nonzeros
of this matrix is possible within the constraints of available memory. This poses
a logistical challenge for the solution of linear systems described by this matrix:
the LU-factorisation of a sparse matrix described in Sect. A.2.1.1 does not result
in sparse matrices L and U, and so these matrices will suffer from the memory

limitations described earlier. To circumvent this problem, iterative techniques may be used for the solution of sparse linear systems, where successive iterates of the solution of the linear system $\mathbf{x}_k$, $k = 1, 2, \ldots$ are generated until $\|\mathbf{b} - \mathbf{A}\mathbf{x}_k\| < \varepsilon$ for some user-specified tolerance ε. This branch of numerical linear algebra is a large subject in its own right and we only touch briefly upon it here, giving one algorithm for a very specific class of matrices, namely symmetric, positive definite matrices.

Algorithm 1 Conjugate gradient method for solving $\mathbf{A}\mathbf{x} = \mathbf{b}$

Require: Symmetric, positive definite matrix $\mathbf{A}$, specified vector $\mathbf{b}$, initial guess $\mathbf{x}_0$ (or set $\mathbf{x}_0 = \mathbf{0}$), tolerance ε.

1: $k = 0$, $\mathbf{r} = \mathbf{b} - \mathbf{A}\mathbf{x}_k$, $\mathbf{p} = \mathbf{0}$, $\beta = 0$
2: **while** $\|\mathbf{r}\| \geq \varepsilon$ **do**
3:   **if** $k > 0$ **then**
4:     $\beta = \dfrac{\mathbf{r}^\top \mathbf{r}}{\mathbf{r}_{\text{prev}}^\top \mathbf{r}_{\text{prev}}}$
5:   **end if**
6:   $\mathbf{p} = \mathbf{r} + \beta\mathbf{p}$
7:   $\alpha = \dfrac{\mathbf{r}^\top \mathbf{r}}{\mathbf{p}^\top \mathbf{A}\mathbf{p}}$
8:   $\mathbf{x}_{k+1} = \mathbf{x}_k + \alpha\mathbf{p}$
9:   $\mathbf{r}_{\text{prev}} = \mathbf{r}$
10:   $\mathbf{r} = \mathbf{b} - \mathbf{A}\mathbf{x}_{k+1}$
11:   $k = k + 1$
12: **end while**
13: $\mathbf{x} = \mathbf{x}_k$

A matrix $\mathbf{A}$ is said to be *positive definite* if, and only if, for all vectors $\mathbf{x}$ of the correct size the following two conditions are met:

$$\mathbf{x}^\top \mathbf{A}\mathbf{x} \geq 0, \quad \text{and}$$
$$\mathbf{x}^\top \mathbf{A}\mathbf{x} = 0, \quad \text{only if } \mathbf{x} = \mathbf{0}.$$

If a matrix $\mathbf{A}$ is positive definite and symmetric, then we may solve the linear system using the *conjugate gradient method*, given by Algorithm 1.

Other Programming Constructs You Might Meet

Below we briefly describe some programming constructs that other programmers may include in their C++ code. Many of these are constructs that were originally designed for the C programming language. As C++ was developed from C, much of the C language is legal C++, although the modifications developed for the C++ language are generally superior.

B.1 C Style Output

We devoted the whole of Chap. 3 to describing the C++ machinery for input and output. To explain the corresponding machinery in C would require a similar amount of space, and so we only touch upon C style output here, limiting ourselves to describing output to the console. Nevertheless, this should give the flavour of C style output commands, allowing the reader to at least recognise them should they see them.

In the code below, we show how to use C style output to print a double precision floating point variable to the screen in both normal and scientific notation, and how to print an integer to the screen. C style output requires the whole of the output to be enclosed within double quotation marks. When a variable is to be printed it is represented by `%f` for a double precision floating point variable, `%i` for an integer variable, and `%e` for a double precision floating point variable in scientific notation. Finally, the variables to be printed are included in an ordered list at the end of the statement. Note that the included file for C style printing is `<stdio.h>`—standard input and output which provides basic functionality similar to `<iostream>` in C++.

© Springer International Publishing AG, part of Springer Nature 2017
J. Pitt-Francis and J. Whiteley, *Guide to Scientific Computing
in C++*, Undergraduate Topics in Computer Science,
https://doi.org/10.1007/978-3-319-73132-2

```
1   #include <stdio.h>
2
3   int main(int argc, char* argv[])
4   {
5      double x = 105.0;
6      int j = 500;
7      printf("x = %f and j = %i\n", x, j);
8      printf("In scientific notation, x = %e\n", x);
9      return 0;
10  }
```

Other C variations on printf which you might meet are fprintf for printing to file, in which the first argument is a file pointer of type FILE* and sprintf for printing to a string.

B.2 C Style Dynamic Memory Allocation

In Sect. 4.2 we explained how the C++ keywords new and delete could be used to allocate memory dynamically for arrays, and then free the memory when i was no longer needed. C also allows this, through the use of malloc ("memory allocate") and free. As with C style output above, we only touch briefly on the use of these functions to allow the reader to recognise them should they come across them. In the code below, we declare a pointer to a double precision variable, vector in line 6. In line 7, we then use the malloc function to allocate memory for 100 entries of the array vector, all of the same size as a double precision floating point variable. In lines 8–13, we use these entries in the same way as a C++ array. Finally in line 14, we free the memory allocated to this array through the use of the C function called free.

```
1   #include <iostream>
2   #include <cstdlib>
3
4   int main(int argc, char* argv[])
5   {
6      double* vector;
7      vector = ((double*)(malloc(100*sizeof(double))));
8      vector[0] = 1.0;
9      vector[90] = 3.0;
10     std::cout << "Entry of vector with index 0 = "
11               << vector[0] << "\n";
12     std::cout << "Entry of vector with index 90 = "
13               << vector[90] << "\n";
14     free(vector);
15     return 0;
16  }
```

B.3 Ternary ?: Operator

In Sect. 2.1.3 we saw that the keywords `if` and `else` could be used to execute one set of statements if a condition was met, and a different set of instructions if the condition is not met, as in the code fragment below.

```
double a, b, x;
if (a > b)
{
    x = 100.0;
}
else
{
    // a <= b
    x = 0.0;
}
```

The ternary[3] ?: operator has identical effect to the `if-else` statements above: the code above may be written identically as

```
double a, b, x;
x = (a > b) ? 100.0 : 0.0;
```

Although the code written above is shorter than the original `if-else` statements we do not recommend it. The use of `if` and `else` makes the code much more readable, especially by anyone who is not an expert in C++ programming.

B.4 Using Namespace

You may find it tedious to have to write `std::` before `cout` and other functionality of the C++ language. There is a way around this—we may use the `using` statement once in the code as shown below.

```
#include <iostream>

using namespace std;
int main(int argc, char* argv[])
{
    string city = "Cambridge";
    cout << city << "\n";
    return 0;
}
```

---

[3] A ternary operator has three inputs.

At first sight, the code above may appear to make a programmer's life a little easier. Both `string` and `cout` have been used here without being preceded by the slightly clunky `std::`. This approach does, however, introduce a subtle problem. Suppose we declared a variable called "`vector`". It would then be unclear whether an instance of the word "`vector`" is referring to this variable, or the STL vector introduced in Chap. 8, which the `using` statement now allows us to refer to as `vector` rather than `std::vector`. As such, we do not recommend use of the `using` keyword.

B.5 Structures

A *structure* is a collection of variables that are combined together. Structures can be thought of as very simple classes, but without the ability to declare functions, access privileges, or any other properties of classes other than variables. An example of a structure is shown below. Note how the variables are accessed in exactly the same way as classes (using "." for a member or "->" to access a member by de-referencing a pointer).

```cpp
#include <iostream>

struct ModelParameters
{
    double viscosity;
    double density;
    int numberOfDimensions;
};

int main(int argc, char* argv[])
{
    ModelParameters example1;
    example1.viscosity = 1.0e-4;
    example1.density = 1.0;
    ModelParameters* p_eg1 = &example1;
    p_eg1->numberOfDimensions = 3;

    std::cout << "Density is " << example1.density << "\n";

    return 0;
}
```

B.6   Multiple Inheritance

As mentioned in Sect. 7.1 C++, unlike many other object-oriented languages, allows *multiple inheritance* in which a derived class can be derived from multiple base classes. That is, classes may have more than one parent.

Suppose we require a class of matrices so that we can calculate the determinant of given matrices, calculate the eigenvalues of these matrices, and calculate the norm of these matrices. One colleague may have a class of matrices, MatrixDet, that calculates the determinant of a matrix, but doesn't have the functionality for calculating the eigenvalues or the norm of a matrix. Another colleague may have a class of matrices, MatrixEigsNorm, that does allow us to calculate the eigenvalues and norm of a matrix, but not the determinant. The functionality required is therefore all available, but not in the same class. It would therefore be convenient to merge the two classes to create a new class that contains all the functionality required. This is possible through *multiple inheritance*. Below we show how to perform multiple inheritance to generate a new class MatrixCombined.

```
1  #include "MatrixDet.hpp"
2  #include "MatrixEigsNorm.hpp"
3
4  class MatrixCombined: public MatrixDet,
5                        public MatrixEigsNorm
6  {
7      // Body of class
8  };
```

If the class MatrixDet has no member with the same name as a member of the class MatrixEigsNorm then multiple inheritance is an ideal solution to this problem. Suppose both classes have a method called ZeroEntries. Provided this member is made a virtual function in both the class MatrixDet and the class MatrixEigsNorm we may prevent ambiguity through either defining a new function in the class MatrixCombined, or by explicitly identifying which function is to be used in the calling code. For example:

```
1      MatrixCombined mat;
2      // use method ZeroEntries from the class MatrixDet
3      mat.MatrixDet::ZeroEntries();
```

B.7   Class Initialisers

In many cases, the constructor of a class is a simple piece of code involving a list of assignments. For example, the default constructor for the Book class in

Sect. 6.2.7 set all the string fields to "unspecified" and the default constructor o: the ComplexNumber class in Sect. 6.4 set the real and imaginary components t(zero.

```cpp
#include "ComplexNumber.hpp"
// Override default constructor
// Set real and imaginary parts to zero
ComplexNumber::ComplexNumber()
{
   mRealPart = 0.0;
   mImaginaryPart = 0.0;
}
```

In cases where a constructor makes assignments it is more efficient to use C++ *initialisers*. These are comma-separated lists of member variables and values which appear after the constructor's signature (and a colon) but before the main body of the constructor code. Compilers for C++ are able to optimise a list of initialised values more completely than a block of code containing assignment statements. It mus be noted that some compilers insist that the initialisers are ordered exactly as they appear in the definition of the class. An example constructor for the class of complex numbers given in Sect. 6.4 that uses class initialisers is shown below.

```cpp
#include "ComplexNumber.hpp"
// Override default constructor
// Initialize real and imaginary parts as zero
ComplexNumber::ComplexNumber() :
              mRealPart(0.0),
              mImaginaryPart(0.0)
{
   // possibly have more code in body
}
```

Solutions to Exercises

C.1 Matrix and Linear System Classes

The code below is example solutions for the `Matrix` and `LinearSystem` classes developed in the Exercises at the end of Chap. 10.

Listing C.1 `Matrix.hpp`

```
1   #ifndef MATRIXHEADERDEF
2   #define MATRIXHEADERDEF
3   #include "Vector.hpp"
4
5   class Matrix
6   {
7   private:
8      double** mData; // entries of matrix
9      int mNumRows, mNumCols; // dimensions
10  public:
11     Matrix(const Matrix& otherMatrix);
12     Matrix(int numRows, int numCols);
13     ~Matrix();
14     int GetNumberOfRows() const;
15     int GetNumberOfColumns() const;
16     double& operator()(int i, int j); //1-based indexing
17     //overloaded assignment operator
18     Matrix& operator=(const Matrix& otherMatrix);
19     Matrix operator+() const; // unary +
20     Matrix operator-() const; // unary -
21     Matrix operator+(const Matrix& m1) const; // binary +
22     Matrix operator-(const Matrix& m1) const; // binary -
23     // scalar multiplication
```

© Springer International Publishing AG, part of Springer Nature 2017
J. Pitt-Francis and J. Whiteley, *Guide to Scientific Computing in C++*, Undergraduate Topics in Computer Science,
https://doi.org/10.1007/978-3-319-73132-2

```
24      Matrix operator*(double a) const;
25      double CalculateDeterminant() const;
26      // declare vector multiplication friendship
27      friend Vector operator*(const Matrix& m,
28                                const Vector& v);
29      friend Vector operator*(const Vector& v,
30                                const Matrix& m);
31   };
32   // prototype signatures for friend operators
33   Vector operator*(const Matrix& m, const Vector& v);
34   Vector operator*(const Vector& v, const Matrix& m);
35
36   #endif
```

Listing C.2 Matrix.cpp

```
1    #include <cmath>
2    #include <cassert>
3    #include "Matrix.hpp"
4    #include "Vector.hpp"
5
6
7    // Copy constructor
8    // Allocate memory for new matrix, and copy
9    // entries into this matrix
10   Matrix::Matrix(const Matrix& otherMatrix)
11   {
12      mNumRows = otherMatrix.mNumRows;
13      mNumCols = otherMatrix.mNumCols;
14      mData = new double* [mNumRows];
15      for (int i=0; i<mNumRows; i++)
16      {
17         mData[i] = new double [mNumCols];
18      }
19      for (int i=0; i<mNumRows; i++)
20      {
21         for (int j=0; j<mNumCols; j++)
22         {
23            mData[i][j] = otherMatrix.mData[i][j];
24         }
25      }
26   }
27
28   // Constructor for vector of a given length
29   // Allocates memory, and initialises entries
30   // to zero
31   Matrix::Matrix(int numRows, int numCols)
32   {
33      assert(numRows > 0);
34      assert(numCols > 0);
35      mNumRows = numRows;
```

```cpp
36       mNumCols = numCols;
37       mData = new double* [mNumRows];
38       for (int i=0; i<mNumRows; i++)
39       {
40          mData[i] = new double [mNumCols];
41       }
42       for (int i=0; i<mNumRows; i++)
43       {
44          for (int j=0; j<mNumCols; j++)
45          {
46             mData[i][j] = 0.0;
47          }
48       }
49    }
50
51    // Overwritten destructor to correctly free memory
52    Matrix::~Matrix()
53    {
54       for (int i=0; i<mNumRows; i++)
55       {
56          delete[] mData[i];
57       }
58       delete[] mData;
59    }
60
61    // Method to get number of rows of matrix
62    int Matrix::GetNumberOfRows() const
63    {
64       return mNumRows;
65    }
66
67    // Method to get number of columns of matrix
68    int Matrix::GetNumberOfColumns() const
69    {
70       return mNumCols;
71    }
72
73    // Overloading the round brackets
74    // Note that this uses 'one-based' indexing,
75    // and a check on the validity of the index
76    double& Matrix::operator()(int i, int j)
77    {
78       assert(i > 0);
79       assert(i < mNumRows+1);
80       assert(j > 0);
81       assert(j < mNumCols+1);
82       return mData[i-1][j-1];
83    }
84
85    // Overloading the assignment operator
86    Matrix& Matrix::operator=(const Matrix& otherMatrix)
```

```
87   {
88      assert(mNumRows = otherMatrix.mNumRows);
89      assert(mNumCols = otherMatrix.mNumCols);
90
91      for (int i=0; i<mNumRows; i++)
92      {
93         for (int j=0; j<mNumCols; j++)
94         {
95            mData[i][j] = otherMatrix.mData[i][j];
96         }
97      }
98      return *this;
99   }
100
101  // Overloading the unary + operator
102  Matrix Matrix::operator+() const
103  {
104     Matrix mat(mNumRows, mNumCols);
105     for (int i=0; i<mNumRows; i++)
106     {
107        for (int j=0; j<mNumCols; j++)
108        {
109           mat(i+1,j+1) = mData[i][j];
110        }
111     }
112     return mat;
113  }
114
115  // Overloading the unary - operator
116  Matrix Matrix::operator-() const
117  {
118     Matrix mat(mNumRows, mNumCols);
119     for (int i=0; i<mNumRows; i++)
120     {
121        for (int j=0; j<mNumCols; j++)
122        {
123           mat(i+1,j+1) = -mData[i][j];
124        }
125     }
126     return mat;
127  }
128
129  // Overloading the binary + operator
130  Matrix Matrix::operator+(const Matrix& m1) const
131  {
132     assert(mNumRows == m1.mNumRows);
133     assert(mNumCols == m1.mNumCols);
134     Matrix mat(mNumRows, mNumCols);
135     for (int i=0; i<mNumRows; i++)
136     {
137        for (int j=0; j<mNumCols; j++)
```

```
138            {
139                mat(i+1,j+1) = mData[i][j] + m1.mData[i][j];
140            }
141        }
142        return mat;
143  }
144
145  // Overloading the binary - operator
146  Matrix Matrix::operator-(const Matrix& m1) const
147  {
148        assert(mNumRows == m1.mNumRows);
149        assert(mNumCols == m1.mNumCols);
150        Matrix mat(mNumRows, mNumCols);
151        for (int i=0; i<mNumRows; i++)
152        {
153            for (int j=0; j<mNumCols; j++)
154            {
155                mat(i+1,j+1) = mData[i][j] - m1.mData[i][j];
156            }
157        }
158        return mat;
159  }
160
161  // Overloading scalar multiplication
162  Matrix Matrix::operator*(double a) const
163  {
164        Matrix mat(mNumRows, mNumCols);
165        for (int i=0; i<mNumRows; i++)
166        {
167            for (int j=0; j<mNumCols; j++)
168            {
169                mat(i+1,j+1) = a*mData[i][j];
170            }
171        }
172        return mat;
173  }
174
175  // Overloading matrix multiplied by a vector
176  Vector operator*(const Matrix& m, const Vector& v)
177  {
178        int original_vector_size = v.GetSize();
179        assert(m.GetNumberOfColumns() == original_vector_size);
180        int new_vector_length = m.GetNumberOfRows();
181        Vector new_vector(new_vector_length);
182
183        for (int i=0; i<new_vector_length; i++)
184        {
185            for (int j=0; j<original_vector_size; j++)
186            {
187                new_vector[i] += m.mData[i][j]*v.Read(j);
188            }
```

```
189        }
190
191        return new_vector;
192    }
193
194    // Overloading vector multiplied by a matrix
195    Vector operator*(const Vector& v, const Matrix& m)
196    {
197        int original_vector_size = v.GetSize();
198        assert(m.GetNumberOfRows() == original_vector_size);
199        int new_vector_length = m.GetNumberOfColumns();
200        Vector new_vector(new_vector_length);
201
202        for (int i=0; i<new_vector_length; i++)
203        {
204            for (int j=0; j<original_vector_size; j++)
205            {
206                new_vector[i] += v.Read(j)*m.mData[j][i];
207            }
208        }
209
210        return new_vector;
211    }
212
213    // Calculate determinant of square matrix recursively
214    double Matrix::CalculateDeterminant() const
215    {
216        assert(mNumRows == mNumCols);
217        double determinant = 0.0;
218
219        if (mNumRows == 1)
220        {
221            determinant = mData[0][0];
222        }
223        else
224        {
225            // More than one entry of matrix
226            for (int i_outer=0; i_outer<mNumRows; i_outer++)
227            {
228                Matrix sub_matrix(mNumRows-1,
229                                  mNumRows-1);
230                for (int i=0; i<mNumRows-1; i++)
231                {
232                    for (int j=0; j<i_outer; j++)
233                    {
234                        sub_matrix(i+1,j+1) = mData[i+1][j];
235                    }
236                    for (int j=i_outer; j<mNumRows-1; j++)
237                    {
238                        sub_matrix(i+1,j+1) = mData[i+1][j+1];
239                    }
```

```
240        }
241        double sub_matrix_determinant =
242                sub_matrix.CalculateDeterminant();
243
244        determinant += pow(-1.0, i_outer)*
245                mData[0][i_outer]*sub_matrix_determinant;
246      }
247    }
248    return determinant;
249 }
```

Listing C.3 LinearSystem.hpp

```
1  #ifndef LINEARSYSTEMHEADERDEF
2  #define LINEARSYSTEMHEADERDEF
3  #include "Vector.hpp"
4  #include "Matrix.hpp"
5
6  class LinearSystem
7  {
8  protected://private or protected for Exercise 10.5
9     int mSize; // size of linear system
10    Matrix* mpA;  // matrix for linear system
11    Vector* mpb;  // vector for linear system
12
13    // Only allow constructor that specifies matrix/vector
14    // to be used.  Copy constructor is private or protected.
15    LinearSystem(const LinearSystem& otherLinearSystem){};
16  public:
17    LinearSystem(const Matrix& A, const Vector& b);
18
19    // destructor frees memory allocated
20    ~LinearSystem();
21
22    // Method for solving system
23    virtual Vector Solve();
24  };
25
26  #endif
```

Listing C.4 `LinearSystem.cpp`

```cpp
#include <cmath>
#include <cassert>
#include "LinearSystem.hpp"
#include "Matrix.hpp"
#include "Vector.hpp"

// Copy matrix and vector so that original matrix and vector
// specified are unchanged by Gaussian elimination
LinearSystem::LinearSystem(const Matrix& A, const Vector& b)
{
   // check matrix and vector are of compatible sizes
   int local_size = A.GetNumberOfRows();
   assert(A.GetNumberOfColumns() == local_size);
   assert(b.GetSize() == local_size);

   // set variables for linear system
   mSize = local_size;
   mpA = new Matrix(A);
   mpb = new Vector(b);
}

// Destructor to free memory
LinearSystem::~LinearSystem()
{
   delete mpA;
   delete mpb;
}

// Solve linear system using Gaussian elimination
// This method changes the content of the matrix mpA
Vector LinearSystem::Solve()
{
   Vector m(mSize); //See description in Appendix A
   Vector solution(mSize);

   // We introduce references to make the syntax readable
   Matrix& rA = *mpA;
   Vector& rb = *mpb;

   // forward sweep of Gaussian elimination
   for (int k=0; k<mSize-1; k++)
   {
      // see if pivoting is necessary
      double max = 0.0;
      int row = -1;
      for (int i=k; i<mSize; i++)
      {
         if (fabs(rA(i+1,k+1)) > max)
         {
            row = i;
```

```
51          max=fabs(rA(i+1,k+1));
52        }
53      }
54      assert(row >= 0);
55
56      // pivot if necessary
57      if (row != k)
58      {
59          // swap matrix rows k+1 with row+1
60          for (int i=0; i<mSize; i++)
61          {
62              double temp = rA(k+1,i+1);
63              rA(k+1,i+1) = rA(row+1,i+1);
64              rA(row+1,i+1) = temp;
65          }
66          // swap vector entries k+1 with row+1
67          double temp = rb(k+1);
68          rb(k+1) = rb(row+1);
69          rb(row+1) = temp;
70      }
71
72      // create zeros in lower part of column k
73      for (int i=k+1; i<mSize; i++)
74      {
75          m(i+1) = rA(i+1,k+1)/rA(k+1,k+1);
76          for (int j=k; j<mSize; j++)
77          {
78              rA(i+1,j+1) -= rA(k+1,j+1)*m(i+1);
79          }
80          rb(i+1) -= rb(k+1)*m(i+1);
81      }
82  }
83
84  // back substitution
85  for (int i=mSize-1; i>-1; i--)
86  {
87      solution(i+1) = rb(i+1);
88      for (int j=i+1; j<mSize; j++)
89      {
90          solution(i+1) -= rA(i+1,j+1)*solution(j+1);
91      }
92      solution(i+1) /= rA(i+1,i+1);
93  }
94
95  return solution;
96 }
```

Listing C.5 LinearSystemTestSuite.hpp

```
1   #include <cmath>
2   #include <cxxtest/TestSuite.h>
3   #include "Vector.hpp"
4   #include "Matrix.hpp"
5   #include "LinearSystem.hpp"
6
7   // An outline solution for Exercise 10.4
8   class LinearSystemTestSuite : public CxxTest::TestSuite
9   {
10  public:
11      // Test constructors (using norm etc.)
12      void TestDefaultConstructors(void)
13      {
14          Matrix squ(5,5);
15          TS_ASSERT_DELTA(squ.CalculateDeterminant(), 0.0, 1e-8);
16          Matrix nonsquare(7,13);
17          TS_ASSERT_EQUALS(nonsquare.GetNumberOfRows(),    7);
18          TS_ASSERT_EQUALS(nonsquare.GetNumberOfColumns(), 13);
19          Vector vec(25);
20          TS_ASSERT_DELTA(vec.CalculateNorm(), 0.0, 1.0e-8);
21          TS_ASSERT_EQUALS(vec.GetSize(), 25);
22          TS_ASSERT_EQUALS(length(vec), 25);
23      }
24      // Empty test
25      void TestSomeExceptions(void)
26      {
27          // Our code uses assertions for error checking.
28          // If you use Exception then test:
29          //      Matrix a(3,3); Matrix b(4,4);
30          //      TS_ASSERT_THROWS_ANYTHING(a+b);
31      }
32      // Test with cond(a) ~= 1e7
33      void TestLargeConditionNumber(void)
34      {
35          Matrix a(3,3);  Vector b(3); Vector x(3);
36          a(1,1) = 1;   a(1,2) = 0;   a(1,3) = 1e7;
37          a(2,1) = 1;   a(2,2) =-1;   a(2,3) = 0;
38          a(3,1) = 1;   a(3,2) = 0;   a(3,3) = 1;
39          b(1) = 1e7+1; b(2) = 0;     b(3) = 2;
40          double det = a.CalculateDeterminant();
41          TS_ASSERT_DELTA(det, 1.0e7-1.0, 1e-8);
42          LinearSystem ls(a, b);
43          x = ls.Solve();
44          for (int i=1; i<=3; i++)
45          {
46              TS_ASSERT_DELTA( x(i), 1.0, 1e-8);
47          }
48      }
49      // Gaussian Elimination without pivoting would fail:
50      void TestZeroPivot(void)
```

```
51    {
52        Matrix a(3,3);  Vector b(3); Vector x(3);
53        a(1,1) = 0;    a(1,2) = 1;    a(1,3) = 1;
54        a(2,1) = 1;    a(2,2) =-1;    a(2,3) = 0;
55        a(3,1) = 1;    a(3,2) = 1;    a(3,3) = 1;
56        b(1) = 2;      b(2) = 0;      b(3) = 3;
57        TS_ASSERT_DELTA( a.CalculateDeterminant(), 1.0, 1e-8);
58        LinearSystem ls(a, b);
59        x = ls.Solve();
60        for (int i=1; i<=3; i++)
61        {
62            TS_ASSERT_DELTA( x(i), 1.0, 1e-8);
63        }
64        TS_ASSERT_DELTA( x.CalculateNorm(1), 3.0,        1e-8);
65        TS_ASSERT_DELTA( x.CalculateNorm(2), sqrt(3.0), 1e-8);
66    }
67 };
```

C.2   ODE Solver Library

The code below is example solutions for the classes developed in the Exercises at the end of Chap. 12.

Listing C.6 FiniteDifferenceGrid.cpp

```
1   #include <cassert>
2   #include "FiniteDifferenceGrid.hpp"
3   #include "Node.hpp"
4
5   FiniteDifferenceGrid::FiniteDifferenceGrid(int numNodes,
6                                   double xMin, double xMax)
7   {
8       double stepsize = (xMax-xMin)/((double)(numNodes-1));
9       for (int i=0; i<numNodes; i++)
10      {
11          Node node;
12          node.coordinate = xMin+i*stepsize;
13          mNodes.push_back(node);
14      }
15      assert(mNodes.size() == numNodes);
16  }
```

Listing C.7 BvpOde.cpp

```cpp
 1  #include <iostream>
 2  #include <fstream>
 3  #include <cassert>
 4  #include "BvpOde.hpp"
 5
 6  BvpOde::BvpOde(SecondOrderOde* pOde,
 7                 BoundaryConditions* pBcs, int numNodes)
 8  {
 9     mpOde = pOde;
10     mpBconds = pBcs;
11
12     mNumNodes = numNodes;
13     mpGrid = new FiniteDifferenceGrid(mNumNodes, pOde->mXmin,
14                        pOde->mXmax);
15
16     mpSolVec = new Vector(mNumNodes);
17     mpRhsVec = new Vector(mNumNodes);
18     mpLhsMat = new Matrix(mNumNodes, mNumNodes);
19
20     mFilename = "ode_output.dat";
21     mpLinearSystem = NULL;
22  }
23
24  BvpOde::~BvpOde()
25  {
26     // Deletes memory allocated in constructor
27     delete mpSolVec;
28     delete mpRhsVec;
29     delete mpLhsMat;
30     delete mpGrid;
31     // Only delete if Solve has been called
32     if (mpLinearSystem)
33     {
34        delete mpLinearSystem;
35     }
36  }
37
38  void BvpOde::Solve()
39  {
40     PopulateMatrix();
41     PopulateVector();
42     ApplyBoundaryConditions();
43     mpLinearSystem = new LinearSystem(*mpLhsMat, *mpRhsVec);
44     *mpSolVec = mpLinearSystem->Solve();
45     WriteSolutionFile();
46  }
47
48  void BvpOde::PopulateMatrix()
49  {
50     for (int i=1; i<mNumNodes-1; i++)
```

```
51    {
52        // xm, x and xp are  x(i-1), x(i) and x(i+1)
53        double xm = mpGrid->mNodes[i-1].coordinate;
54        double x = mpGrid->mNodes[i].coordinate;
55        double xp = mpGrid->mNodes[i+1].coordinate;
56        double alpha = 2.0/(xp-xm)/(x-xm);
57        double beta = -2.0/(xp-x)/(x-xm);
58        double gamma = 2.0/(xp-xm)/(xp-x);
59        (*mpLhsMat)(i+1,i) = (mpOde->mCoeffOfUxx)*alpha -
60                             (mpOde->mCoeffOfUx)/(xp-xm);
61        (*mpLhsMat)(i+1,i+1) = (mpOde->mCoeffOfUxx)*beta +
62                             mpOde->mCoeffOfU;
63        (*mpLhsMat)(i+1,i+2) = (mpOde->mCoeffOfUxx)*gamma +
64                             (mpOde->mCoeffOfUx)/(xp-xm);
65    }
66 }
67
68 void BvpOde::PopulateVector()
69 {
70    for (int i=1; i<mNumNodes-1; i++)
71    {
72        double x = mpGrid->mNodes[i].coordinate;
73        (*mpRhsVec)(i+1) = mpOde->mpRhsFunc(x);
74    }
75 }
76
77 void BvpOde::ApplyBoundaryConditions()
78 {
79    bool left_bc_applied = false;
80    bool right_bc_applied = false;
81
82    if (mpBconds->mLhsBcIsDirichlet)
83    {
84        (*mpLhsMat)(1,1) = 1.0;
85        (*mpRhsVec)(1) = mpBconds->mLhsBcValue;
86        left_bc_applied = true;
87    }
88
89    if (mpBconds->mRhsBcIsDirichlet)
90    {
91        (*mpLhsMat)(mNumNodes,mNumNodes) = 1.0;
92        (*mpRhsVec)(mNumNodes) = mpBconds->mRhsBcValue;
93        right_bc_applied = true;
94    }
95
96    if (mpBconds->mLhsBcIsNeumann)
97    {
98        assert(left_bc_applied == false);
99        double h = mpGrid->mNodes[1].coordinate -
100                   mpGrid->mNodes[0].coordinate;
101       (*mpLhsMat)(1,1) = -1.0/h;
```

```
102        (*mpLhsMat) (1,2) = 1.0/h;
103        (*mpRhsVec) (1) = mpBconds->mLhsBcValue;
104        left_bc_applied = true;
105    }
106
107    if (mpBconds->mRhsBcIsNeumann)
108    {
109        assert(right_bc_applied == false);
110        double h = mpGrid->mNodes[mNumNodes-1].coordinate -
111                   mpGrid->mNodes[mNumNodes-2].coordinate;
112        (*mpLhsMat) (mNumNodes,mNumNodes-1) = -1.0/h;
113        (*mpLhsMat) (mNumNodes,mNumNodes) = 1.0/h;
114        (*mpRhsVec) (mNumNodes) = mpBconds->mRhsBcValue;
115        right_bc_applied = true;
116    }
117
118    // Check that boundary conditions have been applied
119    // on both boundaries
120    assert(left_bc_applied);
121    assert(right_bc_applied);
122 }
123
124 void BvpOde::WriteSolutionFile()
125 {
126    std::ofstream output_file(mFilename.c_str());
127    assert(output_file.is_open());
128    for (int i=0; i<mNumNodes; i++)
129    {
130        double x = mpGrid->mNodes[i].coordinate;
131        output_file << x << "   " << (*mpSolVec)(i+1) << "\n";
132    }
133    output_file.flush();
134    output_file.close();
135    std::cout<<"Solution written to "<<mFilename<<"\n";
136 }
```

Listing C.8 BoundaryConditions.cpp

```
1  #include <cassert>
2  #include "BoundaryConditions.hpp"
3
4  BoundaryConditions::BoundaryConditions()
5  {
6     mLhsBcIsDirichlet = false;
7     mRhsBcIsDirichlet = false;
8     mLhsBcIsNeumann = false;
9     mRhsBcIsNeumann = false;
10 }
```

```cpp
void BoundaryConditions::SetLhsDirichletBc(double lhsValue)
{
    assert(!mLhsBcIsNeumann);
    mLhsBcIsDirichlet = true;
    mLhsBcValue = lhsValue;
}

void BoundaryConditions::SetRhsDirichletBc(double rhsValue)
{
    assert(!mRhsBcIsNeumann);
    mRhsBcIsDirichlet = true;
    mRhsBcValue = rhsValue;
}

void BoundaryConditions::
          SetLhsNeumannBc(double lhsDerivValue)
{
    assert(!mLhsBcIsDirichlet);
    mLhsBcIsNeumann = true;
    mLhsBcValue = lhsDerivValue;
}

void BoundaryConditions::
          SetRhsNeumannBc(double rhsDerivValue)
{
    assert(!mRhsBcIsDirichlet);
    mRhsBcIsNeumann = true;
    mRhsBcValue = rhsDerivValue;
}
```

Listing C.9 BvpOdeTestSuite.hpp

```cpp
#include <cxxtest/TestSuite.h>
#include <fstream>
#include "BvpOde.hpp"

double model_prob_1_rhs(double x){return 1.0;}
double model_prob_2_rhs(double x){return 34.0*sin(x);}

// This suite is an example solution to Exercise 12.5
class BvpOdeTestSuite : public CxxTest::TestSuite
{
private:
    void ReadIn(const char* rName, std::vector<double>& ts,
                                   std::vector<double>& vs)
    {
        std::ifstream file(rName);
        double time, value;
        while (!file.eof())
```

```
18            {
19                file >> time >> value;
20                if (file.good())
21                {
22                    ts.push_back(time); vs.push_back(value);
23                }
24            }
25        }
26    public:
27        void TestModelProblem1(void)
28        {
29            SecondOrderOde ode_mp1(-1.0, 0.0, 0.0,
30                                   model_prob_1_rhs,
31                                   0.0, 1.0);
32            BoundaryConditions bc_mp1;
33            bc_mp1.SetLhsDirichletBc(0.0);
34            bc_mp1.SetRhsDirichletBc(0.0);
35
36            BvpOde bvpode_mp1(&ode_mp1, &bc_mp1, 101);
37            bvpode_mp1.SetFilename("model_problem_results1.dat");
38            bvpode_mp1.Solve();
39            std::vector<double> xs, us;
40            ReadIn("model_problem_results1.dat", xs, us);
41            TS_ASSERT_EQUALS(xs.size(), 101u);
42            TS_ASSERT_EQUALS(us.size(), 101u);
43            // Test solution as given in Sec. 12.1.1
44            for (int i=0; i<xs.size(); i++)
45            {
46                TS_ASSERT_DELTA(us[i], xs[i]*(1.0-xs[i])/2.0, 1e-8);
47            }
48        }
49
50        void TestModelProblem2(void)
51        {
52            SecondOrderOde ode_mp2(1.0, 3.0, -4.0,
53                                   model_prob_2_rhs,
54                                   0.0, M_PI);
55            BoundaryConditions bc_mp2;
56            bc_mp2.SetLhsNeumannBc(-5.0);
57            bc_mp2.SetRhsDirichletBc(4.0);
58
59            BvpOde bvpode_mp2(&ode_mp2, &bc_mp2, 1001);
60            bvpode_mp2.SetFilename("model_problem_results2.dat");
61            bvpode_mp2.Solve();
62            std::vector<double> xs, us;
63            ReadIn("model_problem_results2.dat", xs, us);
64            TS_ASSERT_EQUALS(xs.size(), 1001u);
65            TS_ASSERT_EQUALS(us.size(), 1001u);
66            // Test solution as given in Sec. 12.1.1
67            for (int i=0; i<xs.size(); i++)
68            {
```

```
69          double u = (4*exp(xs[i])+exp(-4*xs[i]))/
70                     (4*exp(M_PI)+exp(-4*M_PI))
71                     - 5*sin(xs[i]) - 3*cos(xs[i]);
72          TS_ASSERT_DELTA(us[i], u, 2e-3); // Error ~= delta x
73        }
74      }
75  };
```

Further Reading

In earlier chapters we have touched on a few issues that are beyond the scope of this book. When discussing these issues we have directed the interested reader towards a selection of various resources: these are listed below thematically. For the "Mathematical Methods and Linear Algebra" theme, the most comprehensive reference for the basic material is that written by Kreyszig. The other references given are suitable for more advanced numerical concepts. For the "C++ Programming" theme, the website http://www.cplusplus.com provides extensive practical guidance, whilst the texts listed focus on advanced features of the language. In the "Message–Passing Interface" theme the texts give an accessible tutorial–based overview of MPI-1 and MPI-2, respectively. The differences between these two MPI standards are discussed in Sect. 11.2.

Mathematical Methods and Linear Algebra

1. Iserles, A.: A First Course in the Numerical Analysis of Differential Equations, 2nd edn. Cambridge University Press, Cambridge (2009)
2. Kreyszig, E.: Advanced Engineering Mathematics, 9th edn. Wiley, Inc., New York (2006)
3. Süli, E., Mayers, D.F.: An Introduction to Numerical Analysis. Cambridge University Press, Cambridge (2006)
4. Trefethen, L.N., Bau, D.: Numerical Linear Algebra, Society for Industrial and Applied Mathematics (1997)

© Springer International Publishing AG, part of Springer Nature 2017
J. Pitt-Francis and J. Whiteley, *Guide to Scientific Computing in C++*, Undergraduate Topics in Computer Science,
https://doi.org/10.1007/978-3-319-73132-2

C++ Programming

5. Cline, M.P., Lomow, G., Girou, M.: C++ FAQs, 2nd edn. Addison–Wesley, Boston (1998)
6. Meyers, S.: Effective C++, 3rd edn. Addison–Wesley, Boston (2005)
7. Stroustrup, B.: The C++ Programming Language, 3rd edn. AT&T (2000)
8. The Website, http://www.cplusplus.com

The Message–Passing Interface (MPI)

9. Gropp, W., Lusk, E., Skjellum, A.: Using MPI: Portable Parallel Programming with the Message–Passing Interface, 2nd edn. Massachussetts Institute of Technology Press, Massachussetts (1999)
10. Gropp, W., Lusk, E., Thakur, R.: Using MPI-2: Advanced Features of the Message–Passing Interface. Massachussetts Institute of Technology Press, Massachussetts (1999)

Index

© Springer International Publishing AG, part of Springer Nature 2017   283
J. Pitt-Francis and J. Whiteley, *Guide to Scientific Computing
in C++*, Undergraduate Topics in Computer Science,
https://doi.org/10.1007/978-3-319-73132-2

Printed in the United States
By Bookmasters